AN ANALYTICAL CALCULUS

VOLUME I

AN ANALYTICAL CALCULUS

FOR SCHOOL AND UNIVERSITY

BY

E. A. MAXWELL

Fellow of Queens' College, Cambridge

VOLUME I

CAMBRIDGE
AT THE UNIVERSITY PRESS
1966

CAMBRIDGE UNIVERSITY PRESS
Cambridge, New York, Melbourne, Madrid, Cape Town, Singapore, São Paulo, Delhi

Cambridge University Press
The Edinburgh Building, Cambridge CB2 8RU, UK

Published in the United States of America by Cambridge University Press, New York

www.cambridge.org
Information on this title: www.cambridge.org/9780521056960

First published 1954
Reprinted 1966
This digitally printed version 2008

A catalogue record for this publication is available from the British Library

ISBN 978-0-521-05696-0 hardback
ISBN 978-0-521-09035-3 paperback

DEDICATED
TO
MY CHILDREN

PHYLLIS, SHEILA, DENNIS, JEAN

CONTENTS

Chapter V: DEVICES IN INTEGRATION

Chapter VI: APPLICATIONS OF INTEGRATION

PREFACE

I HAVE been fortunate in the help received during the preparation of this work. The manuscript was read with great thoroughness by Dr Sheila M. Edmonds, of Newnham College, Cambridge, whose criticisms and suggestions were of great value and kept me firmly in the paths of rigour. Dr J. W. S. Cassels, of Trinity College, Cambridge, read the proofs and drew attention to a number of slips. To both I would express my sincere thanks.

A number of pupils helped me in the preparation of the answers. Special mention must be made of Mr J. E. Wallington and Mr P. A. Wallington, who, acting almost as a committee, provided me with a complete set of checked answers; any slips that remain must be due to my own carelessness in transcription. I am deeply indebted to them for a very substantial piece of work.

As on former occasions, I have been greatly helped by the staff of the Cambridge University Press, and I should like to place on record how much I owe to their skilled interpretation of the manuscript.

The Examples come from many sources—the Oxford and Cambridge Schools Examination Board, Scholarship examinations in the University of Cambridge, and Degree Examinations in the Universities of Cambridge and London. I am grateful for permission to reproduce them.

E. A. M.

QUEENS' COLLEGE CAMBRIDGE
June, 1953

NOTE ON THE THIRD IMPRESSION

A number of corrections have been made, mainly small. I am indebted to Mr L. E. Clarke of the University College of Ghana for a very helpful list covering this and the later volumes.

INTRODUCTION

THE AIM of these volumes is that they shall together form a complete course in Calculus from its beginnings up to the point where it joins with the subject usually known as analysis. The whole conception is based on considerable dissatisfaction with much that seems rough-and-ready in the basic ideas with which pupils reach the universities, so that almost anything seems acceptable for 'proof' which is superficially plausible. Of course the early work cannot be treated with the rigour appropriate to more mature judgement; but I have tried here, however unsuccessfully, to present the subject in such a way that the more exact treatment, when it comes, can follow by natural development, without being forced to return to a fresh beginning which is often felt to be both unnecessary and even pointless. (How many students lose the thread of analysis just because they do not see any reason for the first few lectures and therefore do not give them serious attention?)

The first volume deals with the basic ideas of differentiation and integration. Graphical methods are used freely, but, it is hoped, in such a way that the essential logical development is never far away. The examples at this stage are mainly very simple, and beginners should have no difficulty in acquiring a fluent technique. Integration appears from the start as area and summation, the method of calculation by inverse differentiation being deduced. All the usual elementary functions are treated, but the logarithmic and exponential functions are postponed.

In Volume II, more advanced parts of the theory make their appearance. The logarithmic and exponential functions are treated in some detail, followed by Taylor's series and the hyperbolic functions. The treatment of curves seems somewhat different from that usually adopted—for example, the formula $\tan\psi = rd\theta/dr$ is derived without the help of the 'elementary triangle', and other ideas often left to intuition are developed on a logical basis. There is a lengthy account of complex numbers, and the volume concludes with 'infinite' integrals and systematic integration. The examples include many that are simple, but go up to scholarship, or early university, level.

Volume III contains a treatment of the functions of several variables, including both partial differentiation and multiple integration, and also a chapter on curve-tracing. We are now definitely at sixth form, or first-year university, level. Volume IV deals with more advanced work, such as differential equations, Fourier series and similar topics. For these two volumes, the examples are mainly of university standard.

It is hoped that a reader will find material for a continuous study of the subject from the start until he leaves it at the end for more advanced study, or perhaps earlier in order to apply it to other sections of his work. Although I have tried to keep the standard of discussion at a level which the mathematical specialist will appreciate, I have also tried to remember the needs of others and given much attention to keeping the actual exposition as simple and clear as is possible. In particular, I hope that the scientist and the engineer will find all their basic needs in suitably digestible form. But I ought to state explicitly that, although a number of practical applications appear as illustrations, I have made no attempt at all to write a specifically 'applied' calculus. My own feeling is that the underlying foundations should be made firm by a study of calculus in its own right; the applications can then be made by others according to their own individual requirements. To attempt the two things at once may lead to confusion.

CHAPTER I

THE IDEA OF DIFFERENTIATION

1. Functions. If a stone is thrown vertically upwards from the ground, it gets slower and slower till, at a certain height, it stops and immediately begins to fall down again. When the speed of projection is u feet per second, the height s feet at time t seconds is given approximately by the equation

$$s = ut - 16t^2.$$

This formula expresses a relation between the three quantities s, u, t; when two of them are known, the third can be calculated.

(i) If u, t are given, then s is determined uniquely; the height can be calculated at a given time for a definite speed of projection.

(ii) If s, t are given, then u is determined uniquely; the speed of projection can be calculated for a given height at a given time.

(iii) If s, u are given, the equation for t is the *quadratic*

$$16t^2 - ut + s = 0,$$

so that there are two values of t; the particle, projected with given speed, is at a given height twice (provided, of course, that it gets there at all), once going up and once coming down.

We say that there is a *functional relation* connecting s, u, t, or that each of them is a *function* of the other two. The *magnitudes* connected by the relationship are called *variables*; those to which we assign values of our own choosing are called *independent* variables and those whose values are then restricted (perhaps even determined, as in the example quoted) are called *dependent*. Thus the value of a dependent variable is governed by that of the independent variables.

In the example just given, the functional relationship was expressed by means of a precise algebraic formula. It may be helpful to point out at once that variables may be functions of each other even when such a formula is lacking; the essential thing is that there should be *some* rule whereby the dependent variable may be found when the independent is given. Suppose, for example, that y is the first prime number greater than x.

When x is given, y is completely determined, but there is no formula connecting them. All the same, y is a definite function of x.

EXAMPLES I

1. Write down the functional relation connecting x and y for the following data:

(i) The product of x^2 and y exceeds y^3 by 4.

(ii) The sum of x^2 and y^2 exceeds the square root of y by 2.

2. State the functional relation connecting the sides a, b, c of a triangle ABC in which the angle at A is a right angle. Taking a, b as independent variables, express c in terms of them.

Let us return to the formula

$$s = ut - 16t^2.$$

Each of the variables is a function of the other two, but *in a particular problem* any one of them may remain fixed in value. The most natural example would be the motion of a particle with given speed of projection u. Then we should regard u as a *constant* and the formula as a relation between TWO variables s, t, each being a function of the other.

A number such as u, which enters into the formula but remains unchanged throughout the problem, is called a *parameter*. Its presence influences the 'size' of the genuine variables.

There is one difficulty to which we should refer at once. Consider the two functions

$$\frac{1}{x}, \quad \sin\left(\frac{1}{x}\right)^\circ,$$

where the symbol $^\circ$ is used to denote measurement in degrees. Each has a definite value for every value of x, *except when x is zero.*

(i) The function $\frac{1}{x}$ assumes the form $\frac{1}{0}$ when x is zero, but there exists no such number.

(ii) The value of $\sin\left(\frac{1}{x}\right)^\circ$ always lies between -1 and $+1$, but nothing at all can be said when x is zero. We may, for example, evaluate $\sin\left(\frac{1}{x}\right)^\circ$ for $x = \cdot 1, \cdot 01, \cdot 001, \ldots$ (i.e. $\sin 10^\circ, \sin 100^\circ, \sin 1000^\circ, \ldots$), but we are no nearer to a definite value for $x = 0$.

In what follows, we shall assume that a function has a definite unique value for each relevant value of the independent variable, except perhaps in cases which we shall indicate explicitly.

EXAMPLES II

1. For what values, if any, of the independent variable x have the following functions no definite value?

$$\frac{1}{x-1}, \quad \frac{1}{x^2-1}, \quad \frac{1}{x^2-4x+5}, \quad \frac{1}{x^2-4x+3}.$$

2. For what values of x have the functions

$$\frac{1}{\sin x}, \quad \tan x$$

no definite value?

3. For what values of x have the functions

$$\cos\frac{1}{x}, \quad \sin^2\left(\frac{\pi}{x^2-1}\right)$$

no definite value?

2. Discrete and continuous variables. Consider the following examples for a function y of an independent variable x:

(i) y is defined in terms of x by means of the relation $y = 2x+3$;

(ii) y is defined to be the first positive integer greater than x;

(iii) y is the average height in inches of the first x men named in the Cambridge Telephone Directory for 1952;

(iv) y is the number of children of the xth man named in the Cambridge Telephone Directory for 1952.

There are important differences in the allowable values of the variables. Let us take the examples in order:

(i) The values of x and y are both unrestricted; each can take any value, positive or negative, integral or fractional.

(ii) The value of x is unrestricted, but y must be a positive integer. That is, the value of y moves by 'jumps'.

(iii) The value of x is necessarily a positive integer; the value of y may, by chance, be an integer, but is more likely to be a fraction, probably between 54 and 78.

(iv) The values of x, y are both positive integers, with y possibly zero.

A variable whose values proceed by steps is called *discrete*; a variable which can take all values (at any rate within limits relevant to the problem in hand) is called *continuous*.

3. Notation. We now direct our attention to functions of a single continuous variable. A function of the independent variable x will be denoted by a symbol such as $f(x), g(x), F(x)$, and so on. Thus, if the function were $2x^2-3$, we should write

$$f(x) \equiv 2x^2 - 3,$$

where the sign '$\equiv$' is used to mean 'is identically equal to', or 'stands for'. For example,

$$f(4) \equiv 2.4^2 - 3 = 29, \quad f(0) \equiv 2.0^2 - 3 = -3.$$

When more than one function is considered, we use separate symbols for each; for example, we might have the three functions

$$f(x) \equiv 1 + x^2, \quad g(x) \equiv 2x, \quad h(x) \equiv 1 - x^2.$$

These functions are actually related, being connected by the identity

$$\{f(x)\}^2 \equiv \{g(x)\}^2 + \{h(x)\}^2.$$

It is sometimes convenient to omit the reference to the independent variable when no confusion can arise, and to use the shorter symbols f, g, F, and so on, for the functions. The identity just given then appears more compactly in the form

$$f^2 \equiv g^2 + h^2.$$

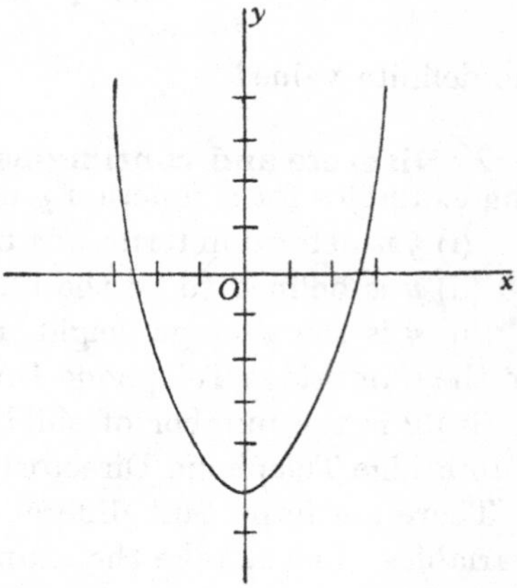

Fig. 1.

In practice, too, we often use a SINGLE LETTER (usually y) to denote a function of the independent variable x. Thus, we might write

$$y \equiv x^2 - 5$$

to mean 'y is the function x^2-5'. By convention, however, the ordinary sign for equality is usually employed in this context, and we write

$$y = x^2 - 5.$$

This notation is linked up with the common representation of a function by means of a *graph*. The diagram (Fig. 1) shows the graph $y = x^2 - 5$ representing the function just quoted.

A function may on occasion receive the alternative names y or $f(x)$ according to convenience, and the graph $y = f(x)$ is then called the *graph of the function*.

We assume that the reader has already had practice in the drawing of graphs. The following revision examples are typical.

EXAMPLES III

Draw the following graphs, choosing your own scales and ranges of values for x:

1. $y = x + 3$.
2. $y = 2x^2 - 3$.
3. $y = \sin x$.
4. $y = x \cos x$.
5. $y = (x-1)(x-2)$.
6. $y = (x-1)(x-2)(x-3)$.
7. $y = x^2(x-1)$.
8.* $y = \sqrt{x}$.

4. Limits. Suppose that a stone, initially at rest, is dropped from some point at a certain height above the ground. As it falls, it is subject to resistance from the air, and some idea of the motion may be found by making the simplifying assumption that the resistance is proportional to the speed—say k times the speed per unit mass. It may then be proved that, if g (approximately equal to 32) is the usual constant of gravitation, the speed after t seconds is

$$\frac{g}{k} - \frac{g}{k e^{kt}},$$

where e is a number about which we shall have much to say later, but which for the present we may take as having a value near to 2·7.

As t increases from zero, e^{kt} increases steadily, and the term $g/(ke^{kt})$ gets less and less, becoming almost negligible for large values of t. Hence as time goes on the speed of the stone (if it has far enough to fall before reaching the ground) approaches more and more closely to the value g/k given by the first term. This value is, in fact, called the *terminal speed* of the stone.

* Here and elsewhere the symbol $\sqrt{}$ implies the *positive* square root unless the contrary is stated explicitly.

It follows that the expression for the speed of the stone is a function of the time which approaches more and more nearly to the value g/k as t becomes larger and larger. We say that the function tends to the *limit* g/k for large values of t.

We have begun by explaining the idea of a limit with the help of a physical example, in which that limit is approached for large values of the independent variable. More generally, we must consider the limit approached by a function as the independent variable tends towards any given *finite* value. In many cases the answer is obvious; for example, the limit of $1+x$ as x approaches 2 is 3. But it is not so obvious what we are to say about $\dfrac{x^2-1}{x-1}$ as x approaches 1, or about $\dfrac{\sin x}{x}$ as x approaches 0. This is the problem which we now begin to examine.

Suppose that $f(x)$ is a given function of x, and that we wish to discuss what happens to $f(x)$ as x tends to a certain value a. Suppose, too, that there exists a number L (which, in practice, may require some skill to determine) such that $f(x)$ is very near to L whenever x is very near to a. We say that L is the *limit* of $f(x)$ as x tends to a, and we write this statement in the form

$$\lim_{x\to a} f(x) = L.$$

We must, of course, have some criterion to apply to the words 'very near'—that is, we must devise a 'test of nearness'. For this purpose we choose any small positive number, which it is customary to call ϵ. We might take ϵ to be ·1, ·01, ·00001, and so on, according to the degree of accuracy which we propose to adopt. The point is that ϵ is an arbitrary number of our own selection, and that it may be taken as small as ever we please. In order to say that $f(x)$ tends to the limit L, we are to insist that the numerical value of the difference between $f(x)$ and L, written

$$|f(x)-L|,$$

is less than this number ϵ for all values of x sufficiently near to a; and this must be true however small ϵ is selected.

We require, in fact, that when ϵ is given there can be found a certain positive number η (whose value will depend on ϵ) such that the difference $|f(x)-L|$ really *is* less than ϵ whenever x differs from a by less than η.

Definition. *A function $f(x)$ tends to the limit L as x tends to the value a if, when ϵ is a given positive number (however small) a positive number η can be found, depending on ϵ, such that*

$$|f(x)-L|<\epsilon$$

whenever $$0<|x-a|<\eta.$$

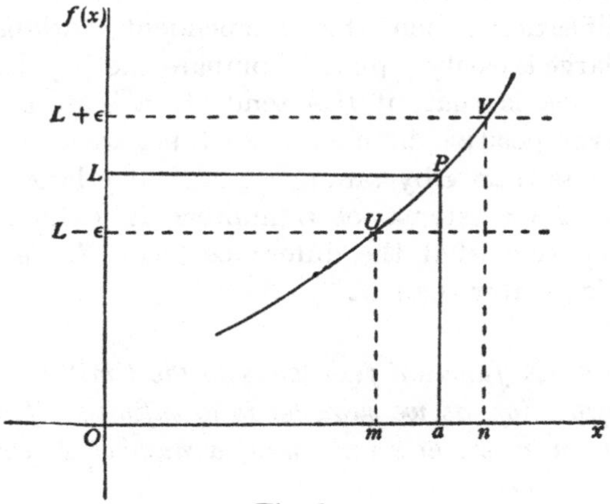

Fig. 2.

The definition may be illustrated graphically. The diagram portrays the graph $y = f(x)$, in which $y = L$ when $x = a$, the corresponding point on the graph being P. The band between the dotted lines at the levels $L-\epsilon, L+\epsilon$ encloses that part of the curve for which $f(x)$ lies between $L-\epsilon, L+\epsilon$; the points where the curve meets the dotted lines are U, V, giving values m, n for x. If we take η to be any number less than both $a-m$ and $n-a$, then the value of $f(x)$ is between $L-\epsilon$ and $L+\epsilon$ whenever x lies between $a-\eta$ and $a+\eta$; that is, $|f(x)-L|$ is less than ϵ whenever $|x-a|$ is less than η. Hence $\lim_{x\to a} f(x) = L$.

Note. The function may never actually TAKE the value towards which it TENDS. For example, the relation

$$\frac{x^2-1}{x-1} = x+1$$

is true whenever x is not equal to 1, and this is so however close to 1 it may be. For example, if $x = 1{\cdot}0000001$, then

$$(x^2-1)/(x-1) = 2{\cdot}0000001.$$

Hence
$$\lim_{x\to 1}\frac{x^2-1}{x-1} = 2.$$

But the function itself has no value when $x = 1$, since numerator and denominator are both zero.

The modification when the independent variable becomes indefinitely large is easily supplied (compare the physical example, p. 5). The idea is that, if $f(x)$ tends to a limit L for large x, then, whatever positive number ϵ we take, we can ensure that $|f(x)-L|$ is less than ϵ by taking x sufficiently large; in fact, we must ensure the existence of a number N (whose value will depend on ϵ), such that the difference $|f(x)-L|$ is less than ϵ whenever x is greater than N.

DEFINITION. *A function $f(x)$ tends to the limit L as x becomes larger and larger* (*or, as we say, 'tends to infinity'*) *if, when ϵ is a given positive number, however small, a number N can be found, depending on ϵ, such that*

$$|f(x)-L| < \epsilon$$

whenever
$$x > N.$$

The symbol '∞' is often used for 'infinity', and the statement

'$f(x)$ tends to L as x tends to infinity'

can be written in the form

$$\text{`}\lim_{x\to\infty} f(x) = L\text{'}.$$

ILLUSTRATION 1. In order to show what is implied by these definitions of a limit, we consider the two examples

$$\lim_{x\to 1}\frac{x+5}{x+2},\quad \lim_{x\to\infty}\frac{x+5}{x+2}.$$

In each case the limiting value itself is easily obtained. We demonstrate how the value fits in with our formal definitions.

(i) We should naturally expect the solution

$$\lim_{x\to 1}\frac{x+5}{x+2} = \frac{1+5}{1+2}$$
$$= 2.$$

For detailed proof, we turn to the definition of a limit, putting $f(x) \equiv (x+5)/(x+2)$, $L = 2$. Then

$$|f(x)-L| = \left|\frac{x+5}{x+2}-2\right|$$
$$= \left|\frac{-x+1}{x+2}\right|.$$

We have to show that, if ϵ is any given positive small number, then we can establish the existence of a positive number η (depending on ϵ) with the property that $\left|\frac{-x+1}{x+2}\right|$ is less than ϵ whenever x lies between $1-\eta$ and $1+\eta$.

Since we are concerned only with values of x near to 1, we may take $x+2$ as positive. The argument divides itself into two parts, according as $x > 1$ or $x < 1$.

If $x > 1$, we write $x = 1+\delta$, where δ is positive. Then

$$|-x+1| = |-(1+\delta)+1| = |-\delta| = \delta,$$

and

$$|x+2| = 3+\delta.$$

The condition is therefore

$$\frac{\delta}{3+\delta} < \epsilon,$$

or

$$\delta < 3\epsilon + \delta\epsilon.$$

[When 'clearing fractions' for an inequality, it is essential to be sure that the denominator is positive. This is the point of the remark that $x+2$ may be taken as positive.]

The inequality is certainly true if we choose δ so that

$$\delta < 3\epsilon.$$

If $x < 1$, we write $x = 1 - \delta'$, where δ' is positive. Then

$$|-x+1| = |-(1-\delta')+1| = |\delta'| = \delta',$$

and $|x+2| = 3 - \delta'$. The condition is therefore

$$\frac{\delta'}{3-\delta'} < \epsilon,$$

or, the denominator being positive,

$$\delta' < 3\epsilon - \delta'\epsilon,$$

or

$$(1+\epsilon)\delta' < 3\epsilon,$$

or

$$\delta' < \frac{3\epsilon}{1+\epsilon}.$$

If we choose δ and δ' so that both are less than $3\epsilon/(1+\epsilon)$, then each of the inequalities

$$\delta < 3\epsilon, \quad \delta' < \frac{3\epsilon}{1+\epsilon}$$

is satisfied. In other words, if we take

$$\eta = \frac{3\epsilon}{1+\epsilon},$$

then the expression $(x+5)/(x+2)$ differs from 2 by less than ϵ for all values of x in the range

$$1 - \eta < x < 1 + \eta.$$

Hence, by the formal definition of a limit,

$$\lim_{x \to 1} \frac{x+5}{x+2} = 2.$$

(ii) Consider next the limit

$$\lim_{x \to \infty} \frac{x+5}{x+2}.$$

We write the expression $(x+5)/(x+2)$ in the form

$$\frac{1 + \left(\frac{5}{x}\right)}{1 + \left(\frac{2}{x}\right)}.$$

As x becomes very large, $5/x$ and $2/x$ become very small, so that

$$\lim_{x\to\infty}\frac{x+5}{x+2}=\frac{1+0}{1+0}$$
$$=1.$$

Now turn to the corresponding formal definition, and put $f(x)=(x+5)/(x+2), L=1$. Then

$$|f(x)-L|=\left|\frac{x+5}{x+2}-1\right|$$
$$=\left|\frac{3}{x+2}\right|.$$

We have to show that, if ϵ is any given positive small number, then we can establish the existence of a number N (depending on ϵ) with the property that $\left|\frac{3}{x+2}\right|$ is less than ϵ whenever x is greater than N.

As we are concerned with large values of x, we need only consider x to be positive, so that

$$\left|\frac{3}{x+2}\right|=\frac{3}{x+2},$$

and the condition is

$$\frac{3}{x+2}<\epsilon,$$

or, the denominator being positive,

$$3<\epsilon x+2\epsilon,$$

or $$\epsilon x>3-2\epsilon,$$

or, finally, $$x>\frac{3-2\epsilon}{\epsilon}.$$

If we write $$N=\frac{3-2\epsilon}{\epsilon},$$

then the expression $(x+5)/(x+2)$ will differ from 1 by less than ϵ for all values of x greater than N. Hence, by the formal definition of a limit,

$$\lim_{x\to\infty}\frac{x+5}{x+2}=1.$$

EXAMPLES IV

Evaluate the following limits:

1. $\lim_{x \to 1} \dfrac{x+5}{x+11}$.
2. $\lim_{x \to \infty} \dfrac{1+x}{1-x}$.
3. $\lim_{x \to 0} \dfrac{x^2+3x+2}{x^2-3x+2}$.
4. $\lim_{x \to \infty} \dfrac{3}{4+5x}$.
5. $\lim_{x \to \infty} \dfrac{5x+7}{3x+5}$.
6. $\lim_{x \to 2} \dfrac{x^2-4}{x-2}$.
7. $\lim_{x \to -1} \dfrac{x^2-1}{x+1}$.
8. $\lim_{x \to 2} \dfrac{x^2-4}{x^2-5x+5}$.

5. Continuity. The treatment of the preceding paragraph must sometimes be modified. Anyone looking at the diagram (Fig. 3) would agree instinctively that the function $f(x)$ represented there is 'continuous' for negative values of x and 'continuous' for positive values of x, but 'discontinuous' when x is zero. We must examine the idea of discontinuity more closely.

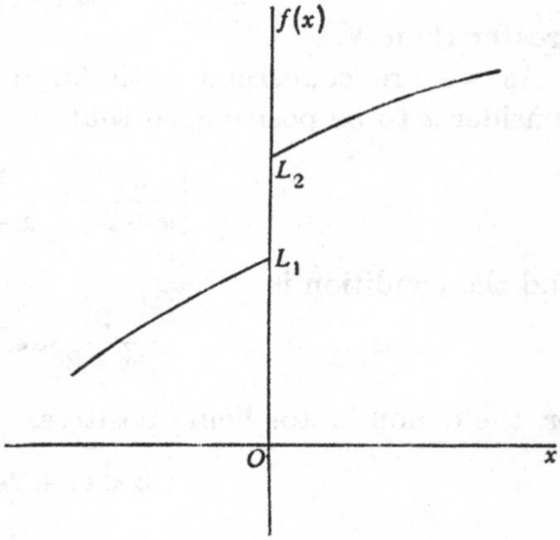

Fig. 3.

We can, if necessary, split up the work of the preceding paragraphs to define TWO limits at $x = a$:

(i) The limit as x tends to a 'from above', where we replace the conditions $0 < |x-a| < \eta$ in the definition by the conditions $0 < x-a < \eta$, making x greater than a. This corresponds to an approach to the point L_2 in the diagram.

(ii) The limit as x tends to a 'from below', where we replace the conditions $0 < |x-a| < \eta$ in the definition by the conditions $0 < a-x < \eta$, making x less than a. This corresponds to an approach to the point L_1 in the diagram.

We shall not have much to say about these two limits, except to note the definition of continuity which they imply:

DEFINITION. *The function $f(x)$ is* CONTINUOUS *when $x = a$ if $f(x)$ tends to a limit L as x tends to a from above and to the same limit L as x tends to a from below, while $f(x) = L$ when $x = a$.*

Continuity is therefore a LOCAL property, as the diagram indicates. Continuity at one point by no means implies continuity at any other.

One of the commonest cases of discontinuity which occurs in practice is when $f(x)$ assumes the form of a 'fraction with a zero denominator'. The simplest illustration is the function

$$f(x) \equiv \frac{1}{x},$$

and the diagram (Fig. 4) shows the graph

$$y = \frac{1}{x}.$$

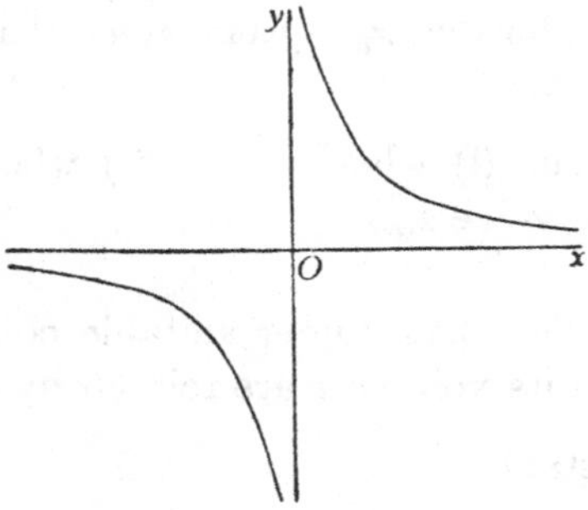

Fig. 4.

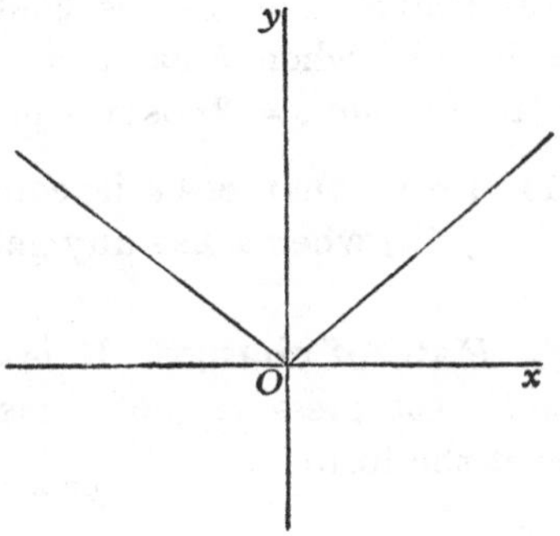

Fig. 5.

For very small values of x, we see that y is large; but y jumps from very large NEGATIVE values to very large POSITIVE values as x moves across the value zero. There is therefore a discontinuity at $x = 0$; indeed, the formula does not define $f(x)$ at all when x is zero.

Similarly a function such as

$$\frac{x+1}{x-1}$$

has a discontinuity at $x = 1$.

We add that a function such as

$$y = |x|$$

(i.e. $y =$ the numerical value of x), whose graph we show (Fig. 5), is NOT discontinuous at $x = 0$.

EXAMPLES V

Name any points at which the following functions are discontinuous:

1. $\frac{1}{x-1}$. 2. $\frac{1}{x^2-1}$. 3. $\frac{x^2-4}{x^2-9}$.

4. $\frac{x}{x^2-3x+2}$. 5. $\frac{1}{\sin x}$. 6. $\frac{1}{\cos x}$.

7. $\tan x$. 8. $\frac{\sin x}{x-1}$. 9. $\frac{\cos x}{x^2-9}$.

10. $\frac{1}{1-\sin x}$. 11. $\frac{1+x^2}{4-\cos x}$. 12. $\frac{1+x^3}{1+2\cos x}$.

13. Prove that $\sin x$ is continuous (i) when $x = 0$, (ii) when $x = \frac{1}{3}\pi$, (iii) when x has any general value x_0. [Remember that $\sin(x+h) - \sin x = 2\cos(x+\frac{1}{2}h)\sin\frac{1}{2}h$.]

14. Prove that $\cos x$ is continuous (i) when $x = 0$, (ii) when $x = \frac{1}{2}\pi$, (iii) when x has any general value x_0.

6. Rate of change. It is familiar that, under suitable conditions, the pressure p of a gas and its volume v are related by a law of the form

$$pv = \text{constant},$$

say

$$pv = b,$$

so that p is the function of v given by the relation

$$p = \frac{b}{v}.$$

Suppose now that the volume of the gas is altered slightly; the pressure will change in sympathy. A standard notation is to denote the small change in the value of v by the symbol

$$\delta v \quad \text{[read 'delta } v\text{']}$$

which may, of course, be positive or negative. As a result of this variation, the pressure suffers a change, which we denote similarly by the symbol δp.

[The signs 'δv' and 'δp' are composite symbols for these changes and stand for single algebraic variables. The 'δ', 'v' and 'p' are not separable entities.]

It is an obvious use of ordinary language to think of the ratio

$$\frac{\delta p}{\delta v}$$

as measuring the rate at which p varies with v.

We now form an estimate for this rate when v has a certain given value which we may conveniently denote by v_1. When the volume has been increased by the amount δv_1 (positive or negative) to the value $v_1 + \delta v_1$, the pressure may be written as $p_1 + \delta p_1$, where p_1 is the value of p corresponding to v_1. These numbers are connected by the relation

$$p_1 + \delta p_1 = \frac{b}{v_1 + \delta v_1}.$$

But $$p_1 = \frac{b}{v_1},$$

so that, by subtraction,

$$\delta p_1 = \frac{b}{v_1 + \delta v_1} - \frac{b}{v_1}$$

$$= \frac{-b\delta v_1}{v_1(v_1 + \delta v_1)}.$$

Hence $$\frac{\delta p_1}{\delta v_1} = -\frac{b}{v_1(v_1 + \delta v_1)}.$$

The rate of change of p with v is obviously best measured by taking the values of the ratio $\delta p/\delta v$ for very small changes in v. In fact, if the ratio $\delta p/\delta v$ tends to a limit as the variation δv is taken smaller and smaller, that limit will suit admirably to define the rate of change. Now, v_1 itself being given, we have the result

$$\lim_{\delta v_1 \to 0} \left\{ -\frac{b}{v_1(v_1 + \delta v_1)} \right\} = -\frac{b}{v_1^2},$$

and so the rate of change of pressure, at an instant when the volume is v_1, is measured by the formula

$$-\frac{b}{v_1^2}.$$

Note. Since b is essentially positive, the rate of change is negative. This agrees with the common-sense observation that p decreases as v increases.

7. Gradient. Closely related to the idea of the rate of change of a function is that of the gradient of a curve. We choose an alternative illustration.

The electrical resistance of a metal may be expected to vary with the temperature, and experiment shows that for platinum the resistance R when the temperature is $\theta°$ C. is given by the relation

$$R = R_0(1+\alpha\theta+\beta\theta^2),$$

where R_0 is the resistance at 0° C. and α, β are positive constants.

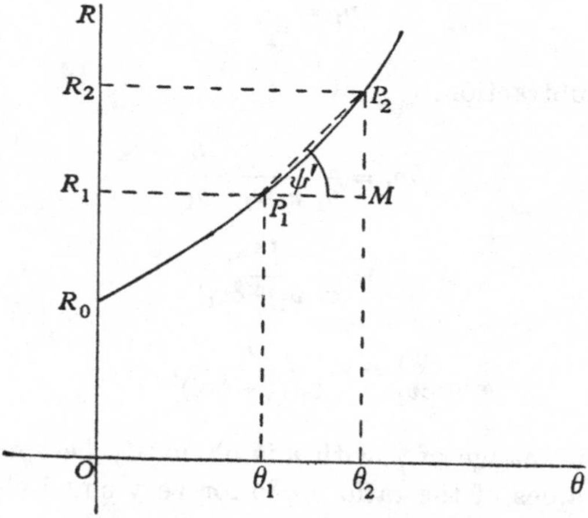

Fig. 6.

The relation between R and θ may be illustrated by means of the accompanying diagram (Fig. 6).

Suppose that the resistance is R_1, R_2 when the temperature is θ_1, θ_2, the corresponding points of the graph being P_1, P_2. The variation of resistance with temperature may suitably be measured by considering the ratio

$$\frac{\text{resistance increase}}{\text{temperature increase}} = \frac{R_2 - R_1}{\theta_2 - \theta_1}.$$

In the diagram, draw the straight line P_1P_2, and also the line P_1M perpendicular to the ordinate through P_2. Then, if the line P_1P_2 makes an angle ψ' with the axis of θ, the value of $\tan\psi'$, being the ratio $(R_2 - R_1)/(\theta_2 - \theta_1)$ just indicated, gives a measure of the rate of change of R with θ *for the range of temperatures* θ_1 *to* θ_2.

It is important to have an expression for the rate of change of R with θ *at the value* θ_1 *itself*. To do this, we take P_2 progressively nearer to P_1; the chord P_1P_2 tends to take up a limiting position called the *tangent* to the curve at P_1, and ψ' assumes a limiting value ψ.

DEFINITION. *The rate of change of R with θ at the point θ_1 is measured by* $\tan\psi$, *which is called the* GRADIENT *of the curve at* P_1.

We can find an expression for the gradient by a method similar to that used in the preceding paragraph. Suppose that the increase of θ from θ_1 to $\theta_1 + \delta\theta_1$ causes an increase of resistance from R_1 to $R_1 + \delta R_1$. Then, by the formula for R,

$$R_1 + \delta R_1 = R_0\{1 + \alpha(\theta_1 + \delta\theta_1) + \beta(\theta_1 + \delta\theta_1)^2\},$$

$$R_1 = R_0\{1 + \alpha\theta_1 + \beta\theta_1^2\}.$$

Subtracting, we have

$$\delta R_1 = R_0\{\alpha\delta\theta_1 + 2\beta\theta_1\delta\theta_1 + \beta(\delta\theta_1)^2\}.$$

Now

$$R_2 - R_1 = \delta R_1,$$

$$\theta_2 - \theta_1 = \delta\theta_1,$$

so that

$$\tan\psi' = \frac{\delta R_1}{\delta\theta_1}$$

$$= R_0\{\alpha + 2\beta\theta_1 + \beta\delta\theta_1\}.$$

By definition, the gradient is

$$\tan\psi = \lim_{\delta\theta_1 \to 0}\left\{\frac{\delta R_1}{\delta\theta_1}\right\}$$

$$= \lim_{\delta\theta_1 \to 0} R_0\{\alpha + 2\beta\theta_1 + \beta\delta\theta_1\}$$

$$= R_0(\alpha + 2\beta\theta_1).$$

Hence *the gradient of the graph* (*the rate of change of* R *with* θ) *at the value* θ_1 *is*

$$R_0(\alpha+2\beta\theta_1).$$

More generally, suppose that

$$y=f(x)$$

is a given curve (Fig. 7) and P, P' two close points upon it given

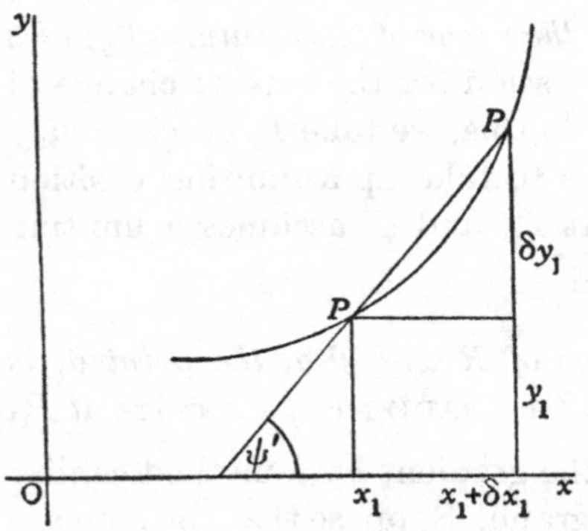

Fig. 7.

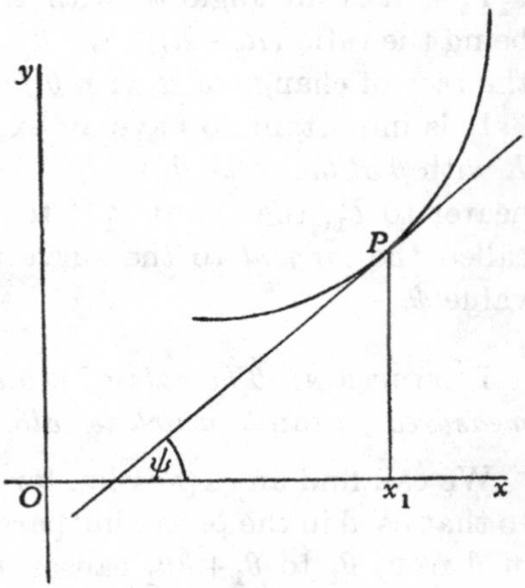

Fig. 8.

by $x=x_1, x_1+\delta x_1$ and $y=y_1, y_1+\delta y_1$ respectively. Then the chord PP' makes an angle ψ' with the x-axis, where

$$\tan\psi' = \frac{\delta y_1}{\delta x_1}$$

$$= \frac{f(x_1+\delta x_1)-f(x_1)}{\delta x_1}.$$

If P' is taken progressively nearer to P (Fig. 8), the chord PP' assumes (in ordinary cases) a limiting position, the *tangent* at P, and ψ' assumes a limiting value ψ, where

$$\tan\psi = \lim_{\delta x_1\to 0}\frac{\delta y_1}{\delta x_1}$$

$$= \lim_{\delta x_1\to 0}\frac{f(x_1+\delta x_1)-f(x_1)}{\delta x_1}.$$

This limiting value of $\tan\psi$ is called the *gradient* of the curve $y=f(x)$ at the point for which x has the value x_1.

EXAMPLES VI

1. Find the gradient of the straight line $y = 3x + 4$ at the points where $x = -2, 0, 2, 4$ respectively.

2. Sketch the curve $y = x^2$, and find the gradient at the points where $x = -1, 0, 1, 2$ respectively.

3. Prove that the gradient of the curve $y = x^3$, at the point where $x = x_1$, is $3x_1^2$.

4. Prove that the gradient of the curve $y = x^2 + 5x + 4$, at the point where $x = x_1$, is $2x_1 + 5$.

5. Prove that the gradient of the curve $y = x^3 - x$, at the point where $x = x_1$, is $3x_1^2 - 1$.

6. Determine the range of values of x for which the gradient of the curve $y = 2x^3 - 9x^2 + 12x$ is negative.

7. Prove that the gradient of the curve $y = x^2 - 4$ is positive for all positive values of x.

8. Find the equation of the tangent to the curve $y = x^2$ at each of the points $(1, 1)$, $(-2, 4)$, $(0, 0)$, $(3, 9)$.

9. Find the equation of the tangent to the curve $y = x^2 + 2x$ at each of the points $(1, 3)$, $(-1, -1)$, $(0, 0)$.

8. The differential coefficient. We now gather together (with some repetition) the ideas of the last two paragraphs.

Let $f(x)$ be a given function, and x_1 a certain value of the independent variable x. Suppose that x_1 receives a small increment; in the notation of the preceding paragraphs this would be called δx_1, but it is now more convenient to use the single letter h. The values of $f(x)$ corresponding to the values $x_1, x_1 + h$ of the independent variable are $f(x_1), f(x_1 + h)$, and the rate of change of $f(x)$ *between the values* $x_1, x_1 + h$ is

$$\frac{f(x_1 + h) - f(x_1)}{h}.$$

The rate of change of $f(x)$ *at the point* x_1 is thus

$$\lim_{h \to 0} \frac{f(x_1 + h) - f(x_1)}{h},$$

and this limit (if it exists) is called *the differential coefficient of* $f(x)$ *with respect to* x *at the value* x_1. The phrase *derivative of* $f(x)$ is also used.

Thus the differential coefficient of b/v with respect to v at the value v_1 is (p. 16) $-b/v_1^2$, and the differential coefficient of $R_0(1+\alpha\theta+\beta\theta^2)$ with respect to θ at the value θ_1 is (p. 18) $R_0(\alpha+2\beta\theta_1)$.

ILLUSTRATION 2. *To evaluate the differential coefficient of* $3x^2-4x$ *with respect to* x *when* $x=4$.

If $f(x) \equiv 3x^2-4x$,

then
$$f(4+h) = 3(4+h)^2-4(4+h)$$
$$= 3(16+8h+h^2)-4(4+h)$$
$$= 32+20h+3h^2,$$

and
$$f(4) = 3.4^2-4.4 = 32.$$

Hence
$$f(4+h)-f(4) = 20h+3h^2,$$

and
$$\frac{f(4+h)-f(4)}{h} = 20+3h.$$

The differential coefficient is the limiting value of this function as h tends to zero, and so its value is 20.

In practice, it is customary to write the value of the differential coefficient so as to refer to a general value x; thus the limit

$$\lim_{h\to 0}\frac{f(x+h)-f(x)}{h}$$

gives the differential coefficient of $f(x)$ at the point x. But it should always be kept in mind that a definite value of x is implied.

ILLUSTRATION 3. *To find the differential coefficient of* x^2.
The differential coefficient is

$$\lim_{h\to 0}\frac{(x+h)^2-x^2}{h} = \lim_{h\to 0}\frac{2hx+h^2}{h} = \lim_{h\to 0}(2x+h)$$
$$= 2x.$$

Hence the differential coefficient of x^2 is $2x$.

Illustration 4. *The function $x^{\frac{1}{3}}$ has no differential coefficient at the point $x = 0$.*

If

$$f(x) = x^{\frac{1}{3}},$$

then

$$\frac{f(0+h)-f(0)}{h} = \frac{f(h)-f(0)}{h} = \frac{h^{\frac{1}{3}}-0}{h}$$

$$= \frac{1}{h^{\frac{2}{3}}}.$$

As h becomes smaller and smaller, $1/h^{\frac{2}{3}}$ becomes larger and larger, and

$$\lim_{h \to 0} \frac{f(0+h)-f(0)}{h}$$

does not exist.

Illustration 5. *The differential coefficient of $1/x$ is $-1/x^2$.* The differential coefficient is

$$\lim_{h \to 0} \frac{\frac{1}{x+h}-\frac{1}{x}}{h} = \lim_{h \to 0} \frac{x-(x+h)}{hx(x+h)}$$

$$= \lim_{h \to 0} \frac{-h}{hx(x+h)} = -\lim_{h \to 0} \frac{1}{x(x+h)}$$

$$= -\frac{1}{x^2}.$$

It is understood that $x \neq 0$, otherwise there is no differential coefficient—indeed, the function is itself undefined.

9. Notation. Several notations are in use for the differential coefficient.

If $f(x)$ is a given function of x, its differential coefficient with respect to x is denoted by the symbol

$$f'(x).$$

Alternatively, if the function is denoted (as, for example, in graphical work) by the letter y, so that

$$y = f(x),$$

then the differential coefficient of y with respect to x is denoted by the symbol

$$\frac{dy}{dx}.$$

Here we do NOT mean a process of division. The notation is an appeal to the eye based on the fact that, if δy is written for $f(x+\delta x)-f(x)$, then

$$\frac{f(x+\delta x)-f(x)}{\delta x}=\frac{\delta y}{\delta x},$$

and the *limit* of this quotient is $\dfrac{dy}{dx}$.

The notations $$y', \quad \frac{d}{dx}\{f(x)\}$$

are also in common use.

When we wish to specify precisely that the differential coefficient is evaluated at, say, $x = a$, we use one of the forms

$$f'(a), \quad \left(\frac{dy}{dx}\right)_a$$

or other obvious modifications.

The process of finding the differential coefficient is known as *differentiation*, or *differentiating the function.*

In illustration of the notation, consider the function x^2 whose differential coefficient (p. 20) is $2x$.

(i) If $$f(x) = x^2,$$
then $$f'(x) = 2x.$$
(ii) If $$y = x^2,$$
then $$\frac{dy}{dx} = 2x,$$
or $$y' = 2x.$$

If we evaluate the differential coefficient when $x = 5$, then

$$f'(5) = 10, \quad \text{or} \quad \left(\frac{dy}{dx}\right)_5 = 10, \quad \text{or} \quad y'_5 = 10.$$

EXAMPLES VII

1. Prove that, if $f(x) \equiv x^2 + x$, then $f'(x) = 2x + 1$ and that $f'(0) = 1$.

2. Prove that, if $y = 4x^2$, then $\dfrac{dy}{dx} = 8x$.

3. Prove that, if $y = 5x^2$, then $y'_3 = 30$.

4. Prove that, if $f(x) \equiv x^3$, then $f'(x) = 3x^2$.

5. Prove that, if $y = x$, then $y' = 1$.

6. Prove that, if $y = x^3 - 8$, then $\dfrac{dy}{dx} = 3x^2$.

7. Prove that, if $y = 5$, then $\dfrac{dy}{dx} = 0$.

8. Prove that, if $y = 4x + 3$, then $y' = 4$.

9. Evaluate $f'(2)$ for each of the functions $4x, 5x^2, x^3$.

10. Prove that the function $\dfrac{1}{x-1}$ has no differential coefficient when $x = 1$.

10. Tangent and normal. We explained in § 7 (p. 18) what is meant by the *tangent* to the curve

$$y = f(x).$$

If P is the point (x_1, y_1) of the curve (Fig. 9), then the tangent

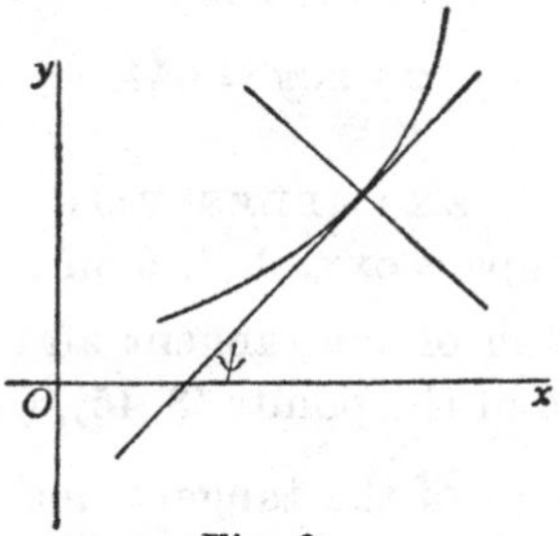

Fig. 9.

at P makes with the x-axis an angle ψ such that

$$\begin{aligned}\tan\psi &= f'(x_1)\\ &= y'_1,\end{aligned}$$

where $f'(x_1) \equiv y_1'$ is the differential coefficient of $f(x)$ evaluated at x_1. Hence, by elementary analytical geometry, the *equation of the tangent at P* is

$$y - y_1 = (x - x_1) f'(x_1),$$

or

$$y - y_1 = (x - x_1) y_1'.$$

DEFINITION. The *normal* to the curve at P is defined to be the line through P perpendicular to the tangent. Hence the *equation of the normal at P* is

$$(y - y_1) f'(x_1) + (x - x_1) = 0,$$

or

$$(y - y_1) y_1' + (x - x_1) = 0.$$

ILLUSTRATION 6. *To find the tangent and normal to the curve*

$$y = 3x^2 - 4x$$

at the point (4, 32).

We have proved (p. 20) that

$$f'(4) = 20.$$

Hence the equation of the tangent is

$$y - 32 = 20(x - 4),$$

or

$$20x - y = 80 - 32$$
$$= 48,$$

and the equation of the normal is

$$20(y - 32) + (x - 4) = 0,$$

or

$$x + 20y = 644.$$

EXAMPLES VIII

[Compare exx. 3, 4, 6 on p. 23.]

1. Find the equation of the tangent and of the normal to the curve $y = 5x^2$ at each of the points (3, 45), (−2, 20), (0, 0).

2. Find the equation of the tangent and of the normal to the curve $y = x^3$ at each of the points (2, 8), (−1, −1), (0, 0).

3. Find the equation of the tangent and of the normal to the curve $y = x^3 - 8$ at the points where it crosses (i) the x-axis, (ii) the y-axis.

CHAPTER II

THE EVALUATION OF DIFFERENTIAL COEFFICIENTS

1. Some theorems on limits. The direct evaluation of a differential coefficient from its definition is often a troublesome matter, and we must now devise a number of rules to shorten our labours. First of all, we 'borrow' from Pure Mathematics some theorems about limits which seem obviously true but which are not too easy to prove rigorously.

(i) *The limit of the sum of two functions is the sum of their individual limits.* Thus

$$\lim_{x\to a}\{f(x)+g(x)\} = \lim_{x\to a} f(x) + \lim_{x\to a} g(x).$$

This result can be extended to any number of functions. For example, we can prove that

$$\lim_{x\to 1}\frac{1-x^2}{1-x} = 2, \quad \lim_{x\to 1}\frac{1-x^3}{1-x} = 3, \quad \lim_{x\to 1}\frac{1-x^4}{1-x} = 4;$$

and so
$$\lim_{x\to 1}\frac{(1-x^2)+(1-x^3)+(1-x^4)}{1-x} = 9.$$

(ii) *The limit of the product of two functions is the product of their individual limits.* Thus

$$\lim_{x\to a}\{f(x)g(x)\} = \lim_{x\to a} f(x) \lim_{x\to a} g(x).$$

This result can be extended to any number of functions.

Properties (i), (ii) can be combined in obvious ways. For example,

$$\lim_{x\to 1}\frac{(1-x^4)(1-x^3)+(1-x^3)(1-x^2)+(1-x^2)(1-x^4)}{(1-x)^2}$$
$$= 4.3+3.2+2.4 = 26.$$

(iii) *The limit of the quotient of two functions is the quotient of their limits,* PROVIDED *that the limit of the denominator is not zero.* Thus

$$\lim_{x\to a}\frac{f(x)}{g(x)}=\frac{\lim\limits_{x\to a} f(x)}{\lim\limits_{x\to a} g(x)},$$

provided that $$\lim_{x\to a} g(x)\neq 0.$$

For example,

$$\lim_{x\to 1}\frac{1-x^4}{1-x^3}=\lim_{x\to 1}\frac{(1-x^4)/(1-x)}{(1-x^3)/(1-x)}=\frac{4}{3}.$$

Our next step is to establish three general theorems, after which we shall derive some standard formulæ for differential coefficients.

2. The differential coefficient of a sum of functions. It is an immediate consequence of § 1 (i), p. 25, that

$$\frac{d}{dx}\{f(x)+g(x)+h(x)+\ldots\}=f'(x)+g'(x)+h'(x)+\ldots,$$

so that *the differential coefficient of the sum of a number of functions is equal to the sum of their differential coefficients.*

The proof is similar to, but easier than, the corresponding theorem for a product of functions given in the following paragraph, and is therefore left as an exercise for the reader.

3. The differential coefficient of the product of two functions. We prove that, *if u, v are two functions of x, then*

$$\frac{d}{dx}(uv)=u\frac{dv}{dx}+v\frac{du}{dx},$$

or (in other notation) $$(uv)'=uv'+vu',$$

provided, of course, that the limits of $\frac{\delta u}{\delta x}, \frac{\delta v}{\delta x}$ exist.

If u, v are the values of the functions when the independent variable has the value x, and $u+\delta u, v+\delta v$ the values for $x+\delta x$, then, by elementary algebra,

$$\frac{(u+\delta u)(v+\delta v)-uv}{\delta x}=u\frac{\delta v}{\delta x}+v\frac{\delta u}{\delta x}+\frac{\delta u\,\delta v}{\delta x}.$$

Since the limits $\lim_{\delta x \to 0} \frac{\delta u}{\delta x}, \quad \lim_{\delta x \to 0} \frac{\delta v}{\delta x}$

exist,

$$\frac{d}{dx}(uv) = \lim_{\delta x \to 0} \frac{(u+\delta u)(v+\delta v) - uv}{\delta x} \qquad \text{[definition; p. 19]}$$

$$= \lim_{\delta x \to 0}\left(u\frac{\delta v}{\delta x}\right) + \lim_{\delta x \to 0}\left(v\frac{\delta u}{\delta x}\right) + \lim_{\delta x \to 0}\left(\frac{\delta u}{\delta x}\delta v\right)$$

[limit of sum; p. 25]

$$= u\frac{dv}{dx} + v\frac{du}{dx} + \frac{du}{dx}\lim_{\delta x \to 0} \delta v \qquad \text{[limit of product; p. 25]}$$

$$= u\frac{dv}{dx} + v\frac{du}{dx},$$

since $\lim_{\delta x \to 0} \delta v$ must be zero for $\lim_{\delta x \to 0} (\delta v/\delta x)$ to exist.

The theorem is therefore proved.

Corollary. By repeated application of this theorem, we have

$$\begin{aligned}(uvw)' &= (uv)'w + (uv)w' \\ &= (u'v + uv')w + (uv)w' \\ &= u'vw + uv'w + uvw'.\end{aligned}$$

In the same way, for any number of functions $u, v, w, \ldots$,

$$(uvw\ldots)' = u'vw\ldots + uv'w\ldots + uvw'\ldots + \ldots.$$

4. The differential coefficient of a 'function of a function'. We prove that, *if $u \equiv f(x)$ is a given function of x, and $y \equiv g(u)$ a given function of u, so that y can be expressed as a function of x in the form $y \equiv g\{f(x)\}$, then*

$$\frac{dy}{dx} = \frac{dy}{du}\frac{du}{dx}.$$

Suppose that when x assumes the value $x + \delta x$, the value of u becomes $u + \delta u$, so that the value of y is $y + \delta y$. Then (but see the *Note* below)

$$\frac{\delta y}{\delta x} = \frac{\delta y}{\delta u}\frac{\delta u}{\delta x},$$

so that

$$\frac{dy}{dx} = \lim_{\delta x \to 0} \frac{\delta y}{\delta x} \qquad \text{[definition; p. 19]}$$

$$= \lim_{\delta x \to 0} \frac{\delta y}{\delta u}\frac{\delta u}{\delta x}$$

$$= \lim_{\delta x \to 0} \frac{\delta y}{\delta u} \lim_{\delta x \to 0} \frac{\delta u}{\delta x} \qquad \text{[product of limits; p. 25].}$$

But $\delta u \to 0$ as $\delta x \to 0$ (assuming that $\frac{du}{dx}$ exists, as is implicit in the enunciation). Hence

$$\frac{dy}{dx} = \lim_{\delta u \to 0} \frac{\delta y}{\delta u} \lim_{\delta x \to 0} \frac{\delta u}{\delta x}$$

$$= \frac{dy}{du}\frac{du}{dx}.$$

COROLLARY. *The differential coefficient of the quotient u/v is*

$$\frac{vu' - uv'}{v^2}.$$

Let $$y = u/v = uv^{-1}.$$

Then, as above, $$\frac{dy}{dx} = \frac{du}{dx}v^{-1} + u\frac{d(v^{-1})}{dx}$$

$$= \frac{du}{dx}v^{-1} + u\frac{d(v^{-1})}{dv}\frac{dv}{dx}.$$

But (p. 21) $$\frac{d(v^{-1})}{dv} = -\frac{1}{v^2},$$

so that $$\frac{dy}{dx} = \frac{1}{v}\frac{du}{dx} - \frac{u}{v^2}\frac{dv}{dx} = \frac{vu' - uv'}{v^2}.$$

Note. When we write $$\frac{\delta y}{\delta x} = \frac{\delta y}{\delta u}\frac{\delta u}{\delta x},$$

we have in mind that the increment δu is not zero. But we know that u is the given function $f(x)$, and so there will usually exist (isolated) values of x at which $f'(x) = 0$; that is to say, the limit of $\delta u \div \delta x$ is zero at such points, and so the initial step of dividing δy by δu is open to the suspicion of being division by a zero denominator. To avoid this difficulty, we proceed as follows:

If $\delta u = 0$, we have

$$\frac{\delta y}{\delta x} = \frac{g(u+\delta u) - g(u)}{\delta x} = \frac{g(u) - g(u)}{\delta x}$$

$$= 0,$$

and so

$$\frac{\delta y}{\delta x} \to 0$$

as $\delta x \to 0$ through values such that δu is zero.

But, by hypothesis, we are examining the case when $\dfrac{du}{dx} = 0$, and so we have the two relations

$$\frac{dy}{dx} = 0, \quad \frac{du}{dx} = 0.$$

Hence it is still true that

$$\frac{dy}{dx} = \frac{dy}{du}\frac{du}{dx}.$$

ILLUSTRATION 1. *To find the differential coefficient of*

$$y = (5x+3)^2.$$

Write

$$u = 5x+3,$$

so that

$$\frac{du}{dx} = 5.$$

We have proved (p. 20) that

$$\frac{d}{du}(u^2) = 2u.$$

Hence

$$\frac{dy}{dx} = \frac{dy}{du}\frac{du}{dx}$$

$$= 2(5x+3) \,.\, 5$$

$$= 10(5x+3).$$

5. The differential coefficient of x^n is nx^{n-1}.

(This result is often proved by using the binomial series for $(x+h)^n$. There is, however, the danger of argument in a circle if the proof of that series depends on the result known as Maclaurin's expansion, which comes later in a normal calculus course. There are, of course, other derivations of the series not open to this objection.)

(i) *Suppose that n is a positive integer.* If we assume that, for some definite value of n, we have

$$\frac{d}{dx}(x^n) = nx^{n-1},$$

then we can prove the general result by induction. For

$$\begin{aligned}\frac{d}{dx}(x^{n+1}) &= \frac{d}{dx}(x^n \cdot x) \\ &= x^n\frac{dx}{dx}+x\frac{d(x^n)}{dx} \quad \text{(p. 26)} \\ &= x^n \cdot 1 + x \cdot nx^{n-1} \\ &= (n+1)x^n.\end{aligned}$$

If the result is true for any particular value of n, then it is true for all subsequent values. But it is true for $n = 0$, since

$$\frac{d}{dx}(x^0) = \frac{d}{dx}(1) = 0.$$

Hence it is true for $n = 1$; hence for $n = 2$; hence for $n = 3$; and so on.

(ii) *Suppose that n is a negative integer.* Write $n = -m$, so that m is a positive integer. Then

$$\begin{aligned}\frac{d}{dx}(x^{-m} \cdot x^m) &= \frac{d}{dx}(x^{-m+m}) = \frac{d}{dx}(x^0) = \frac{d}{dx}(1) \\ &= 0.\end{aligned}$$

[The reader may wish to remind himself from a text-book on algebra about the rules for indices. The basic rule is that $x^p \cdot x^q = x^{p+q}$.]

Hence $$x^{-m}\frac{d(x^m)}{dx}+x^m\frac{d(x^{-m})}{dx}=0,$$

or $$x^{-m}.mx^{m-1}+x^m\frac{d(x^{-m})}{dx}=0,$$

or $$mx^{-1}+x^m\frac{d(x^{-m})}{dx}=0,$$

or $$\frac{d(x^{-m})}{dx}=-mx^{-m-1},$$

or, finally, $$\frac{d(x^n)}{dx}=nx^{n-1}.$$

(iii) *Suppose that n is a rational fraction.* Suppose, that is, that n is of the form p/q, where p, q are integers; we can regard q as positive, but allow p to have either sign. Then

$$(x^{p/q})^q=x^p.$$

Differentiating by means of the formula for a 'function of a function' and the rule already proved for integers, we have

$$q(x^{p/q})^{q-1}\frac{d}{dx}(x^{p/q})=px^{p-1},$$

so that $$\frac{d}{dx}(x^{p/q})=\frac{p}{q}x^{(p-1)-p(q-1)/q}$$

$$=\frac{p}{q}x^{p-1-p+p/q}$$

$$=\frac{p}{q}x^{p/q-1}.$$

Hence $$\frac{d}{dx}(x^n)=nx^{n-1}.$$

(iv) *Suppose that n is irrational* (for example, an unending, non-recurring decimal). We propose to regard it as obvious that n may be approximated as closely as we please by a *rational* fraction, so that the theorem is true to as high a degree of accuracy as we desire. But this seemingly innocent remark covers a number of points of considerable difficulty, and a more advanced text-book must be consulted later when details are required.

EXAMPLES I

Find the differential coefficients of the following functions:

1. x^4. 2. $(x+3)^4$. 3. $(2x+3)^4$.

4. $x^2(x^2+1)$. 5. $x^3(x+1)^3$. 6. $(x+1)^2(x+2)^2$.

7. $\dfrac{1}{x}$. 8. $\dfrac{1}{x^4}$. 9. $\dfrac{1}{(x+2)^3}$.

10. $\dfrac{1}{(2x+3)^5}$. 11. $\dfrac{1}{(3x-5)^7}$. 12. $\dfrac{1}{(4x+3)^2}$.

13. $\sqrt{x}$. 14. $x^{\frac{4}{3}}$. 15. $(x+1)\sqrt{x}$.

16. $\sqrt{(2x+3)}$. 17. $(5x+7)^{\frac{4}{5}}$. 18. $x^{-\frac{4}{3}}$.

19. $(x+7)^{-\frac{1}{2}}$. 20. $(x-3)\sqrt{(2x+5)}$. 21. $\dfrac{x}{x+1}$.

22. $\dfrac{x^2}{(2x+3)^2}$. 23. $\dfrac{\sqrt{x}}{(4x-1)^3}$. 24. $\dfrac{(x+1)^{\frac{1}{2}}}{(x-1)^{\frac{1}{3}}}$.

6. The limit as $x \to 0$ of $\sin x/x$ is unity. In the diagram (Fig. 10), AOB is a triangle, right-angled at A, in which OA is of unit length and the magnitude of $\angle BOA$ is x radians, where x is small. (It should be remembered carefully that, in work of this kind, angles are always measured in RADIANS.) The line AC is drawn perpendicular to OB, and an arc of a circle of unit radius, with its centre at O, passes through A and cuts OB in P. Then P lies between B and C, and we propose to regard it as obvious by intuition that

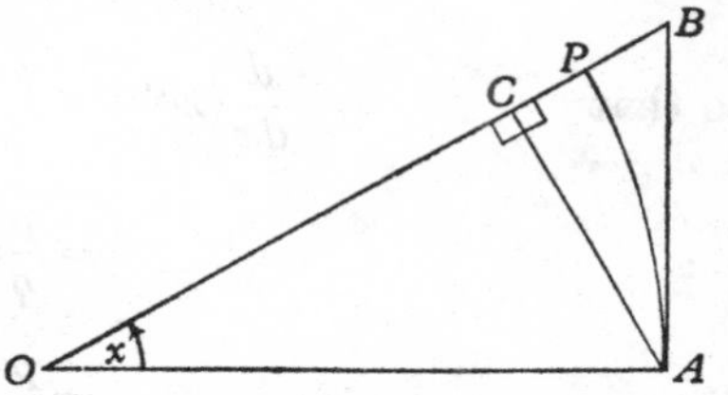

Fig. 10.

$$AC < \text{arc}\, AP < AB.$$

Now $AC = \sin x$, arc $AP = x$, $AB = OB \sin x$, and so

$$\sin x < x < OB \sin x,$$

or
$$1 < \frac{x}{\sin x} < OB.$$

Now let $x \to 0$; then OB tends to equality with OA, so that

$$\lim_{x \to 0} OB = 1.$$

It follows that $\lim_{x \to 0} \frac{x}{\sin x}$, which lies between 1 and $\lim_{x \to 0} OB$, itself has the value 1, and so

$$\lim_{x \to 0} \frac{x}{\sin x} = 1.$$

It is customary to write this result in the inverted form

$$\lim_{x \to 0} \frac{\sin x}{x} = 1.$$

Note. Although $\lim_{x \to 0} \frac{\sin x}{x}$ exists, the value of the function $\frac{\sin x}{x}$ is indeterminate when x is actually zero.

7. The differential coefficient of $\sin x$ is $\cos x$. For

$$\frac{\sin (x+h) - \sin x}{h} = \frac{2 \cos (x + \frac{1}{2}h) \sin \frac{1}{2}h}{h}$$

$$= \cos (x + \tfrac{1}{2}h) . \frac{\sin \frac{1}{2}h}{\frac{1}{2}h}.$$

Now, as $h \to 0$, $\cos (x + \frac{1}{2}h) \to \cos x$

and $\frac{\sin \frac{1}{2}h}{\frac{1}{2}h} \to 1$ (§ 6).

(For the first limit, we have

$$\cos (x + \tfrac{1}{2}h) - \cos x = -2 \sin (x + \tfrac{1}{4}h) \sin \tfrac{1}{4}h.$$

But $\sin (x + \frac{1}{4}h)$ lies between -1 and 1, and $\sin \frac{1}{4}h \to 0$, so that their product also tends to zero. That is,

$$\cos (x + \tfrac{1}{2}h) - \cos x \to 0.)$$

Consequently, since the limit of the product is the product of the limits,

$$\lim_{h \to 0} \frac{\sin (x+h) - \sin x}{h} = \cos x . 1,$$

or

$$\frac{d}{dx} (\sin x) = \cos x.$$

8. The differential coefficient of $\cos x$ is $-\sin x$.

Let $$y = \cos x,$$

and write $$u = \tfrac{1}{2}\pi + x.$$

Then $$y = \cos(u - \tfrac{1}{2}\pi) = \sin u.$$

Hence $$\frac{dy}{du} = \cos u;$$

and also $$\frac{du}{dx} = 1.$$

It follows that
$$\begin{aligned}\frac{dy}{dx} &= \frac{dy}{du}\frac{du}{dx}\\ &= \cos u\\ &= \cos(\tfrac{1}{2}\pi + x)\\ &= -\sin x.\end{aligned}$$

Note. Since $180^\circ = \pi$ radians, the differential coefficient of $\sin x^\circ$ is $(\pi/180)\cos x^\circ$, and the differential coefficient of $\cos x^\circ$ is $-(\pi/180)\sin x^\circ$. [For $\sin x^\circ = \sin(\pi x/180)$ in radian measure.]

EXAMPLES II

(It is most important that the student should acquire complete mastery of the rules so far derived, and examples such as these should be practised regularly.)

Differentiate the following functions:

1. $\sin 2x$. 2. $\cos 3x$. 3. $5\sin 5x$. 4. $x\sin 2x$.
5. $x^2\cos x$. 6. $3x\cos 3x$. 7. $(x+1)^2\sin 7x$. 8. $\sin(3x+5)$.
9. $(2x+1)^3$. 10. $1/(x+2)^4$. 11. $x/\sin x$. 12. $x(1+\sin x)$.
13. $\sin^2 x$. 14. $\cos^2 x$. 15. $\sin^3 x$. 16. $\cos^3 x$.
17. $x\sin^2 x$. 18. $x^2\cos^2 x$. 19. $x^2\cos^2 2x$. 20. $(1+x)\sin^2 x$.
21. $\cos(x-\frac{1}{4}\pi)$. 22. $\sin^2(x+\frac{1}{4}\pi)$. 23. $\sqrt{x}$. 24. $1/\sqrt{x}$.
25. $x^{\frac{3}{2}}\sin^2 x$. 26. $\sqrt[3]{(\sin x)}$. 27. $x\sqrt{(\sin x)}$. 28. $x^2\sqrt{(\sin 4x)}$.
29. $\operatorname{cosec} x$. 30. $\sec x$. 31. $\tan x$. 32. $\cot x$.
33. $\cos 2x^\circ$. 34. $\tan 3x^\circ$. 35. $x\sin x^\circ$. 36. $\sin^2 x^\circ$.

9. Differential coefficients of higher order. If $f(x)$ is a given function of x, its differential coefficient $f'(x)$ is another function of x, having in general its own differential coefficient. This is called the *second differential coefficient of* $f(x)$, and is denoted by the symbol

$$f''(x).$$

Alternatively, if $y = f(x)$, the second differential coefficient of y is written in one or other of the forms

$$\frac{d^2y}{dx^2}, \quad y''.$$

In the same way, the differential coefficient of $f''(x)$ is the third differential coefficient of $f(x)$; and so on. In this way we form a sequence which it is customary to denote by the notations

$$f(x), \quad f'(x), \quad f''(x), \quad f'''(x), \quad f^{(\mathrm{iv})}(x), \quad \ldots, \quad f^{(r)}(x), \quad \ldots$$

$$y, \quad \frac{dy}{dx}, \quad \frac{d^2y}{dx^2}, \quad \frac{d^3y}{dx^3}, \quad \frac{d^4y}{dx^4}, \quad \ldots, \quad \frac{d^ry}{dx^r}, \quad \ldots$$

$$y, \quad y', \quad y'', \quad y''', \quad y^{(\mathrm{iv})}, \quad \ldots, \quad y^{(r)}, \quad \ldots.$$

For example, if

$$y = x^3,$$

then

$$y' = 3x^2, \quad y'' = 6x, \quad y''' = 6, \quad y^{(\mathrm{iv})} = 0.$$

In the case of $\sin x$, $\cos x$, we can obtain a convenient expression for these coefficients:

Let

$$y = \sin x;$$

then

$$y' = \cos x.$$

But, by elementary trigonometry,

$$\cos x = \sin(\tfrac{1}{2}\pi + x),$$

so that

$$y' = \sin(\tfrac{1}{2}\pi + x).$$

Thus the effect of differentiating is to add $\frac{1}{2}\pi$ to the independent variable. Proceeding in this way, we have

$$y'' = \sin(\pi + x),$$

$$y''' = \sin(\tfrac{3}{2}\pi + x),$$

....

$$y^{(r)} = \sin\left(\frac{r}{2}\pi + x\right).$$

In the same way, if

$$y = \cos x,$$

then

$$y^{(r)} = \cos\left(\frac{r}{2}\pi + x\right).$$

EXAMPLES III

Find $\dfrac{dy}{dx}, \dfrac{d^2y}{dx^2}, \dfrac{d^3y}{dx^3}$ for each of the following functions:

1. x^5.
2. x^3.
3. x.
4. $x \sin x$.
5. $x^2 \cos x$.
6. $x \sin^2 x$.
7. $\sin 2x$.
8. $\cos 4x$.
9. $\sin^2 x$.
10. $1/x$.
11. $1/(2x-3)$.
12. $1/x^3$.

13. Prove that, if $f(x)$ is a cubic polynomial in x, then

$$f^{(\text{iv})}(x) = 0.$$

14. Prove that, if $f(x)$ is a polynomial in x of degree n, then $f^{(n)}(x)$ is a constant independent of x.

15. Prove that, if $y = \sin mx$, then

$$\frac{d^2y}{dx^2} + m^2 y = 0.$$

16. Prove that, if $y = x \sin x$, then

$$x^2 y'' - 2xy' + (x^2+2)y = 0.$$

17. Prove that $$\frac{d}{dx}\left(x\frac{dy}{dx}\right) = x\frac{d^2y}{dx^2} + \frac{dy}{dx}.$$

18. Prove that

$$\frac{d^2}{dx^2}\left(x^2\frac{d^2y}{dx^2}\right) = x^2\frac{d^4y}{dx^4} + 4x\frac{d^3y}{dx^3} + 2\frac{d^2y}{dx^2}.$$

10. Some standard forms. There are one or two basic formulæ which follow directly from results already obtained, and which the reader should commit to memory. For convenience, we gather together the standard formulæ of differentiation into this one paragraph.

I. General Rules.

(i) $\frac{d}{dx}(u+v) = \frac{du}{dx}+\frac{dv}{dx}$ (p. 26).

(ii) $\frac{d}{dx}(uv) = u\frac{dv}{dx}+v\frac{du}{dx}$ (p. 26).

(iii) $\frac{d}{dx}\left(\frac{u}{v}\right) = \frac{vu'-uv'}{v^2}$ (p. 28).

(iv) $\frac{dy}{dx} = \frac{dy}{du}\frac{du}{dx}$ (p. 27).

II. Particular Functions.

(i) $\frac{d}{dx}(x^n) = nx^{n-1}$ (p. 30).

(ii) $\frac{d}{dx}(\sin x) = \cos x$ (p. 33).

(iii) $\frac{d}{dx}(\cos x) = -\sin x$ (p. 34).

(iv) $\frac{d}{dx}(\sec x) = \sec x \tan x.$

(v) $\frac{d}{dx}(\tan x) = \sec^2 x.$

(vi) $\frac{d}{dx}(\operatorname{cosec} x) = -\operatorname{cosec} x \cot x.$

(vii) $\frac{d}{dx}(\cot x) = -\operatorname{cosec}^2 x.$

We prove the last four of these results (II, iv–vii):

(iv) Let $y = \sec x.$

Then $y = (\cos x)^{-1},$

so that
$$\frac{dy}{dx} = (-1)(\cos x)^{-2}\frac{d}{dx}(\cos x)$$
$$= (-1)(\cos x)^{-2}(-\sin x) = +\frac{\sin x}{\cos^2 x}$$
$$= \sec x \tan x.$$

(v) Let $$y = \tan x.$$

Then $$y = \sin x \sec x,$$

so that $$\frac{dy}{dx} = \cos x \,.\, \sec x + \sin x \,.\, \sec x \tan x$$
$$= 1 + \tan^2 x$$
$$= \sec^2 x.$$

(vi) Let $$y = \operatorname{cosec} x.$$

Then $$y = (\sin x)^{-1},$$

so that, by reasoning similar to that given for $\sec x$,

$$\frac{dy}{dx} = -(\sin x)^{-2} \cos x$$
$$= -\operatorname{cosec} x \cot x.$$

(vii) Let $$y = \cot x.$$

Then $$y = \cos x \operatorname{cosec} x,$$

so that $$\frac{dy}{dx} = (-\sin x) \operatorname{cosec} x + \cos x \,(-\operatorname{cosec} x \cot x)$$
$$= -1 - \cot^2 x$$
$$= -\operatorname{cosec}^2 x.$$

EXAMPLES IV

Differentiate the following functions:

1. $\sec 2x$.
2. $\sec^2 2x$.
3. $\tan^2 2x$.
4. $\tan^3 3x$.
5. $x \operatorname{cosec} x$.
6. $x^2 \cot 2x$.
7. $\sec x \tan x$.
8. $\sqrt{(\sec x)}$.
9. $\operatorname{cosec}^3 2x$.
10. $x^m \tan^n x$.
11. $x^{-m} \sec^m x$.
12. $x^{\frac{1}{2}} \tan^2 x$.

11.* The inverse circular functions.

I. THE INVERSE SINE. The relation

$$v = \sin u$$

serves to define v in the ordinary way as a function of u. But it can also be used to define u as a function of v, and we use the notation

$$u = \sin^{-1} v$$

* This paragraph may be postponed, if desired.

to mean that u is the function whose sine is v. The notation

$$u = \arcsin v$$

is also used.

It is familiar from elementary trigonometry that, if

$$\sin u = \sin \alpha,$$

then $$u = n\pi + (-1)^n \alpha,$$

where n is a positive or negative integer. Hence the relation $u = \sin^{-1} v$ does not define u as a single-valued function of v. The graph shown in the diagram (Fig. 11) implies this; when u is given, v is determined uniquely, but, when v is given (lying between -1 and $+1$) there are infinitely many values of u.

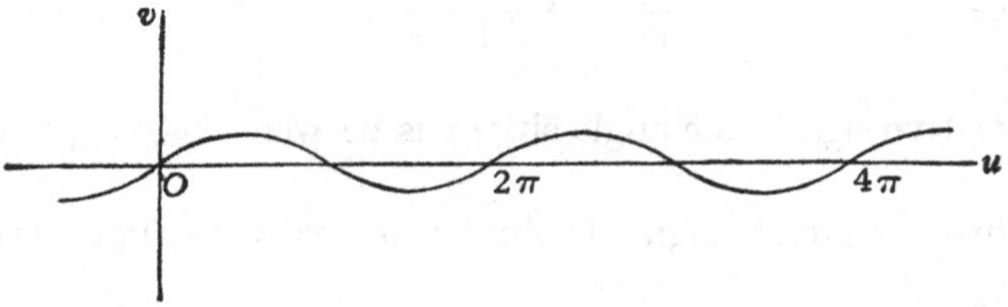

Fig. 11.

Our immediate problem is to evaluate the differential coefficients, with respect to x, of the function $\sin^{-1} x$.

We write

$$y = \sin^{-1} x,$$

so that, by the definition,

$$x = \sin y.$$

Differentiate with respect to x. Then

$$1 = \cos y \frac{dy}{dx},$$

$$\frac{dy}{dx} = \frac{1}{\cos y}$$

$$= \pm \frac{1}{\sqrt{(1-x^2)}}$$

Now inspection of the curve $x = \sin y$, shown in the diagram (Fig. 12), reveals that the gradient $\frac{dy}{dx}$ is *positive* when y lies between $-\frac{\pi}{2}, \frac{\pi}{2}$ and, more generally, between

$$2n\pi - \frac{\pi}{2},\ 2n\pi + \frac{\pi}{2},$$

where n is any positive or negative integer; on the other hand, the gradient is *negative* when y lies between $\frac{\pi}{2}, \frac{3\pi}{2}$ and, more generally, between

$$(2n+1)\pi - \frac{\pi}{2},\ (2n+1)\pi + \frac{\pi}{2}.$$

Hence
$$\frac{dy}{dx} = \pm\frac{1}{\sqrt{(1-x^2)}},$$

with positive sign if the angle $\sin^{-1} x$ is between $2n\pi - \frac{\pi}{2}, 2n\pi + \frac{\pi}{2}$, and with negative sign if $\sin^{-1} x$ is between $(2n+1)\pi - \frac{\pi}{2}$, $(2n+1)\pi + \frac{\pi}{2}$.

In particular, if $\sin^{-1} x$ *is an* ACUTE *angle, then*

$$\frac{dy}{dx} = +\frac{1}{\sqrt{(1-x^2)}}.$$

II. THE INVERSE COSINE. Similar considerations guide us in finding $\frac{dy}{dx}$ when

$$y = \cos^{-1} x,$$

or
$$x = \cos y.$$

Differentiate with respect to x. Then

$$1 = -\sin y \frac{dy}{dx},$$

so that
$$\frac{dy}{dx} = -\frac{1}{\sin y}$$
$$= \pm\frac{1}{\sqrt{(1-x^2)}}.$$

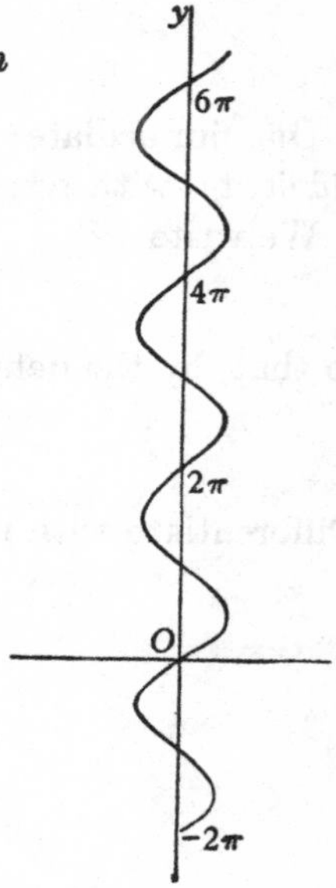

Fig. 12.

To determine the choice of sign, we appeal to the curve $x = \cos y$ shown in the diagram (Fig. 13), and see that the gradient $\frac{dy}{dx}$ is *positive* when y lies between $\pi, 2\pi$ and, more generally, between

$$(2n+1)\pi, (2n+2)\pi;$$

the gradient is *negative* when y lies between $0, \pi$ and, more generally, between

$$2n\pi, (2n+1)\pi.$$

Hence
$$\frac{dy}{dx} = \pm\frac{1}{\sqrt{(1-x^2)}},$$

with positive sign if the angle $\cos^{-1} x$ is between $(2n+1)\pi, (2n+2)\pi$, and with negative sign if $\cos^{-1} x$ is between $2n\pi, (2n+1)\pi$.

In particular, *if* $\cos^{-1} x$ *is an* ACUTE *angle, then*

$$\frac{dy}{dx} = -\frac{1}{\sqrt{(1-x^2)}}.$$

III. THE INVERSE TANGENT. The evaluation of $\frac{dy}{dx}$ when

$$y = \tan^{-1} x,$$

or
$$x = \tan y,$$

is much simpler. Differentiate with respect to x.

Then
$$1 = \sec^2 y \frac{dy}{dx},$$

so that
$$\frac{dy}{dx} = \frac{1}{\sec^2 y}$$

$$= \frac{1}{1+x^2}.$$

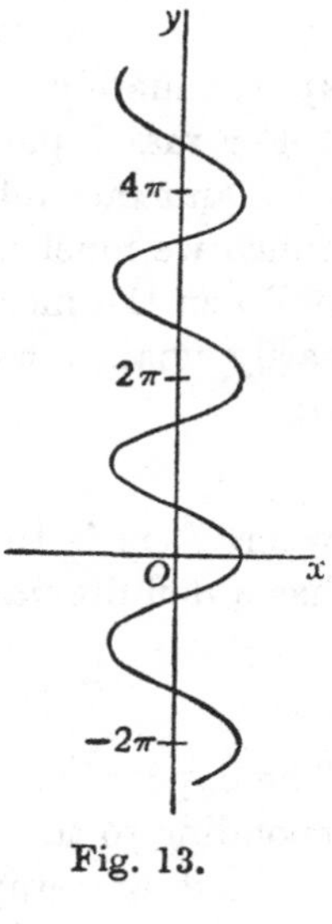

Fig. 13.

EXAMPLES V

1. Evaluate $\frac{dy}{dx}$ for each of the functions $\sin^{-1} x$, $\cos^{-1} x$, $\tan^{-1} x$ when the angle has the value

(i) $\frac{\pi}{3}$, (ii) $\frac{5\pi}{3}$, (iii) $\frac{13\pi}{3}$, (iv) $\frac{5\pi}{6}$, (v) $\frac{7\pi}{6}$, (vi) $\frac{11\pi}{6}$.

Taking the angle to be acute, find $\dfrac{dy}{dx}$ for each of the functions:

2. $x \tan^{-1} x$.
3. $\sin^{-1}(x^2)$.
4. $x \cos^{-1} x$.
5. $x^3 \sin^{-1}(2x)$.
6. $x^2 \cos^{-1}(x^2)$.
7. $(\tan^{-1} x)^2$.
8. $\sec^{-1} x$.
9. $\operatorname{cosec}^{-1} x$.
10. $\cot^{-1} x$.
11. $(\sin^{-1} x)^2$.
12. $x(\cos^{-1} x)^3$.
13. $x^2 \sec^{-1} x$.
14. $x \operatorname{cosec}^{-1} x$.
15. $x^2(\sin^{-1} x)^2$.
16. $1/(\sin^{-1} x)$.

12.* Differentials. It follows, from the definition of the differential coefficient $f'(x)$ of a function $y \equiv f(x)$, that the increment δy in the value of y when x receives a small increment δx is (in ordinary cases) not very different from $f'(x)\,\delta x$, so that

$$\delta y = f'(x)\,\delta x$$

approximately.

For many purposes, however, it is desirable to work instead with an exact relation, and this leads to the idea of a differential, which we must now explain.

When the increment δx is given an arbitrary value, not necessarily small, the expression

$$f'(x)\,\delta x,$$

where $f'(x)$ is the differential coefficient of the function $y \equiv f(x)$, has a definite value, and is denoted by the notation dy, so that

$$dy = f'(x)\,\delta x.$$

The expression dy, or $f'(x)\,\delta x$, is called the *differential* of y corresponding to the increment δx. The relation $dy = f'(x)\,\delta x$ is EXACT, without any approximation; it is an automatic consequence of the definition of the differential. The differential dy and the increment δx are proportional, the coefficient of proportionality being $f'(x)$.

[Note that the increment δy is not usually equal to the differential dy, although they are approximately equal when δx is small.]

* This paragraph may be postponed, if desired.

In particular, we can assign a meaning to the symbol dx, which is the differential of the function x itself corresponding to the increment δx; for the differential coefficient of x is unity, so that

$$\begin{aligned} dx &= 1\,.\,\delta x \\ &= \delta x. \end{aligned}$$

It follows that, if dy, dx are the differentials corresponding to the (arbitrary) increment δx, then

$$dy = f'(x)\,\delta x, \quad dx = \delta x.$$

Hence *the differentials dy, dx satisfy the fundamental exact relation*

$$dy = f'(x)\,dx.$$

The notation dy, dx for differentials is, of course, designed to agree with the composite symbol $\dfrac{dy}{dx}$ for the differential coefficient. In fact, if we divide the relation just given by dx, we obtain the equation

$$dy \div dx = \frac{dy}{dx},$$

or

$$dy = \frac{dy}{dx}\,dx,$$

and obtain $\dfrac{dy}{dx}$ as the actual coefficient of the differential dx.

In practice, especially in physical applications, it is customary to use the symbols dy, dx when the increments $\delta y, \delta x$ are really intended.

With this incorrect usage, the equation

$$dy = f'(x)\,dx$$

is often written where

$$\delta y \simeq f'(x)\,\delta x$$

is meant.

The difference between $dy, \delta y$ may be illustrated graphically, as in the diagram (Fig. 14). Let

$$P(x, y),\, P'(x + \delta x, y + \delta y)$$

be two points of the curve

$$y = f(x).$$

Draw PM perpendicular to the ordinate through P', and let the tangent at P meet that ordinate in Q. Then

$$PM = \delta x = dx,$$

$$MP' = \delta y.$$

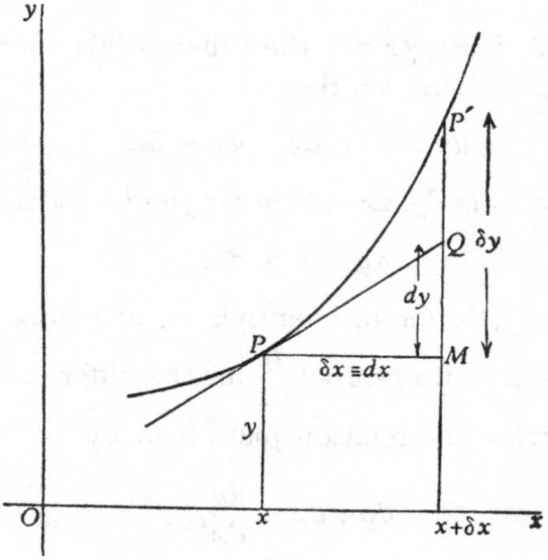

Fig. 14.

Moreover, $$\frac{MQ}{PM} = \tan MPQ$$

$$= \tan(\text{angle between } PQ, Ox)$$

$$= f'(x) \quad \text{(p. 23).}$$

Hence $$MQ = f'(x).PM$$

$$= f'(x)\,\delta x$$

$$= dy,$$

by definition.

Thus MP' represents δy, while MQ represents dy. When δx is small, the difference between MP', MQ is very small indeed; in fact, QP' is then small in comparison with δx.

ILLUSTRATION 2. *A wire is pulled out so that its length is increased by* 1 *per cent. Assuming that the wire can be treated as a cylinder of small cross-section and that the volume remains constant, by what percentage is its diameter decreased?*

Suppose that the constant volume is V and that the length is x. Then the area of a cross-section is V/x, so that the radius r is given by the relation

$$r = \sqrt{\left(\frac{V}{\pi x}\right)} = \sqrt{(V/\pi)}x^{-\frac{1}{2}}.$$

Hence the differentials dr, dx are connected by the relation

$$dr = -\tfrac{1}{2}\sqrt{(V/\pi)}x^{-\frac{3}{2}}\,dx,$$

so that

$$\frac{dr}{r} = -\frac{1}{2}\frac{dx}{x}.$$

For small variations, the differentials are approximately in the same ratio as the increments, so that, if

$$\frac{dx}{x} \simeq \frac{1}{100},$$

then

$$\frac{dr}{r} \simeq -\frac{1}{200},$$

and the decrease in radius (or diameter) is approximately $\frac{1}{2}$ per cent.

EXAMPLES VI

1. If $y = 5x^3$, find the percentage increase in y corresponding to an increase of 2 per cent in x.

2. If $y = 1/x^4$, find the percentage decrease in y corresponding to an increase of $\frac{1}{2}$ per cent in x.

3. If $y = 2x^2$, find the approximate increase in y when x increases from 5 to 5·01.

4. If $y = 1/\sqrt{x}$, find the approximate decrease in y when x increases from 16 to 16·03.

13.* The differentiation of determinants. The process of differentiation when applied to determinants may be illustrated by the special case when the order of the determinant is 4. We then have

$$\Delta \equiv \begin{vmatrix} u_1 & u_2 & u_3 & u_4 \\ v_1 & v_2 & v_3 & v_4 \\ w_1 & w_2 & w_3 & w_4 \\ p_1 & p_2 & p_3 & p_4 \end{vmatrix},$$

* This paragraph may be postponed, if desired.

where $u_1, u_2, \ldots, p_3, p_4$ are all functions of the independent variable x. By definition,

$$\Delta = \Sigma(\pm u_i v_j w_k p_l),$$

where the sign + or − must be taken according as the number of interchanges required to bring the four distinct integers i, j, k, l to the order 1, 2, 3, 4 is even or odd.

It follows that, if dashes denote differentiations with respect to x,

$$\begin{aligned}\Delta' = \Sigma(\pm u_i' v_j w_k p_l) \\ + \Sigma(\pm u_i v_j' w_k p_l) \\ + \Sigma(\pm u_i v_j w_k' p_l) \\ + \Sigma(\pm u_i v_j w_k p_l').\end{aligned}$$

Hence

$$\Delta' = \begin{vmatrix} u_1' & u_2' & u_3' & u_4' \\ v_1 & v_2 & v_3 & v_4 \\ w_1 & w_2 & w_3 & w_4 \\ p_1 & p_2 & p_3 & p_4 \end{vmatrix} + \begin{vmatrix} u_1 & u_2 & u_3 & u_4 \\ v_1' & v_2' & v_3' & v_4' \\ w_1 & w_2 & w_3 & w_4 \\ p_1 & p_2 & p_3 & p_4 \end{vmatrix} + \begin{vmatrix} u_1 & u_2 & u_3 & u_4 \\ v_1 & v_2 & v_3 & v_4 \\ w_1' & w_2' & w_3' & w_4' \\ p_1 & p_2 & p_3 & p_4 \end{vmatrix} + \begin{vmatrix} u_1 & u_2 & u_3 & u_4 \\ v_1 & v_2 & v_3 & v_4 \\ w_1 & w_2 & w_3 & w_4 \\ p_1' & p_2' & p_3' & p_4' \end{vmatrix}.$$

Thus Δ' *is the sum of the four determinants each obtained by differentiating one row of* Δ *and leaving the rest unaltered.*

Similarly, Δ' *is the sum of the four determinants each obtained by differentiating one column of* Δ *and leaving the rest unaltered.*

EXAMPLES VII

1. Prove that, if

$$\Delta \equiv \begin{vmatrix} x^3 & (x+1)^3 & (x+2)^3 \\ x & (x+1) & (x+2) \\ 1 & 1 & 1 \end{vmatrix},$$

then

$$\Delta' = 3\begin{vmatrix} x^2 & (x+1)^2 & (x+2)^2 \\ x & (x+1) & (x+2) \\ 1 & 1 & 1 \end{vmatrix}.$$

2. Prove that, if

$$\Delta \equiv \begin{vmatrix} (x-a)^3 & (x-a)^2 & 1 \\ (x-b)^3 & (x-b)^2 & 1 \\ (x-c)^3 & (x-c)^2 & 1 \end{vmatrix},$$

then

$$\Delta' = 2\begin{vmatrix} (x-a)^3 & x-a & 1 \\ (x-b)^3 & x-b & 1 \\ (x-c)^3 & x-c & 1 \end{vmatrix}.$$

CHAPTER III

APPLICATIONS OF DIFFERENTIATION

1. Illustration from dynamics; velocity, acceleration. Suppose that a particle is moving in a straight line so that its distance at time t from a fixed point O of the line (Fig. 15) is x,

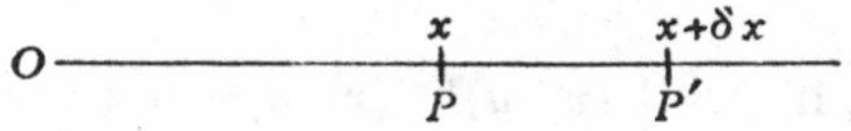

Fig. 15.

and at time $t+\delta t$ is $x+\delta x$. The particle describes the distance δx in the time δt, and so its average speed in that time is

$$\frac{\delta x}{\delta t}.$$

We thus obtain an expression for *the velocity at time t* in the form

$$\lim_{\delta t \to 0} \frac{\delta x}{\delta t}$$

or

$$\frac{dx}{dt}.$$

Let us denote this velocity by u, where

$$u = \frac{dx}{dt},$$

and proceed to find an expression for the rate of change of velocity with time. This rate is, in accordance with our usual principles,

$$\lim_{\delta t \to 0} \frac{\delta u}{\delta t},$$

where $u+\delta u$ is the velocity at time $t+\delta t$. This limit, called the *acceleration* at time t, is

$$\frac{du}{dt},$$

or (p.)

$$\frac{d^2x}{dt^2}.$$

Note. Differentiations with respect to time are often indicated by dots on top of the dependent variable. Thus the velocity and acceleration are $\dot{x}, \ddot{x}$ respectively.

Suppose, for example, that a particle moves in a straight line so that its distance at time t from a fixed origin O is $a \sin nt$; such motion is called *simple harmonic motion.* We have

$$x = a \sin nt,$$

so that
$$\dot{x} = an \cos nt,$$

and
$$\ddot{x} = -an^2 \sin nt$$
$$= -n^2 x.$$

Thus the velocity at time t is $an \cos nt$, and the acceleration is $-an^2 \sin nt$. The acceleration, being also expressed in the form $-n^2x$, is directed TOWARDS the origin and is in magnitude proportional to the distance of the particle from the origin.

EXAMPLES I

Find the velocity and acceleration of a particle moving in a straight line so that its distance x at time t from a fixed point is given by the following relations:

1. $x = \cos \pi t$.
2. $x = \sin \frac{1}{2}\pi t$.
3. $x = 5t - 32t^2$.
4. $x = 32t^2$.
5. $x = t \sin \frac{1}{2}\pi t$.
6. $x = t^2 \cos \frac{1}{2}\pi t$.
7. $x = (t-32)(t-64)$.
8. $x = t \sin^2 \frac{1}{2}\pi t$.

2. Maxima and minima illustrated.

ILLUSTRATION 1. *A log of wood is in the form of a cylinder of radius a. It is required to cut as strong a beam as possible having rectangular section, on the assumption that strength is proportional to width and to the square of height* (Fig. 16).

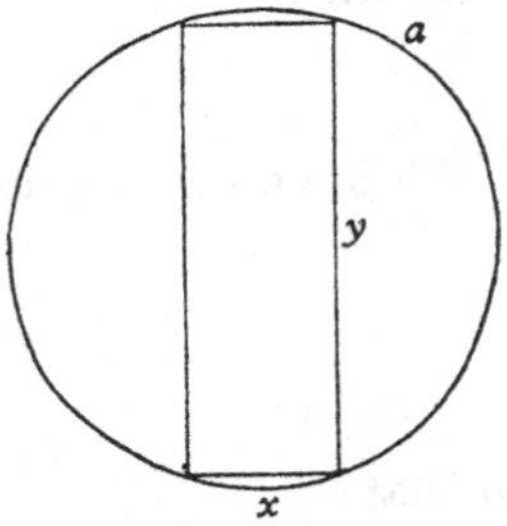

Fig. 16.

If x, y denote the width and height respectively, and s the strength, then

$$s = kxy^2,$$

where k is a constant depending on the nature of the wood.

By the theorem of Pythagoras,

$$x^2+y^2 = 4a^2,$$

so that
$$s = kx(4a^2-x^2).$$

We have to find the greatest value of s.

For this purpose, draw a graph of s against x (Fig. 17). It is easy to obtain the shape shown in the diagram, where, of course, only the part between the values $x = 0, x = 2a$ is relevant.

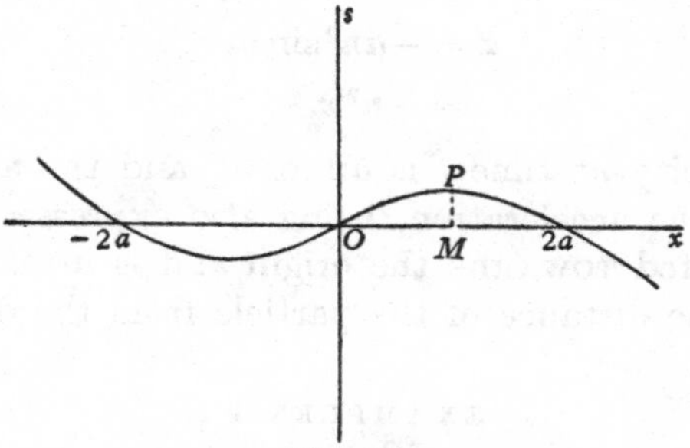

Fig. 17.

A glance at the diagram (Fig. 17) shows that the greatest value of s occurs where $x = OM$ and $s = MP$. Now the characteristic quality of the curve, from which we shall be able to calculate these values of x and s, resides in the fact that the curve has, as it were, 'stopped rising' at P, so that the tangent at P is parallel to Ox. In the language of Chapter I, § 7, p. 16, *the gradient of the curve is zero at P*; that is, *the greatest value of s occurs at a point where*

$$\frac{ds}{dx} = 0.$$

We have the relation

$$s = kx(4a^2-x^2)$$
$$= 4a^2kx-kx^3,$$

so that
$$\frac{ds}{dx} = 4a^2k-3kx^2.$$

Hence $\frac{ds}{dx} = 0$ when
$$x^2 = \tfrac{4}{3}a^2,$$

or
$$x = \pm\tfrac{2}{3}a\sqrt{3}.$$

The relevant value of x is positive, and so we see that the beam of greatest strength has width $\frac{2}{3}a\sqrt{3}$, height $\frac{2}{3}a\sqrt{6}$, and therefore strength $\frac{16}{9}ka^3\sqrt{3}$.

EXAMPLES II

1. Taking $k = \frac{1}{3}$, $a = 1$, draw the graph $s = \frac{1}{3}x(4-x^2)$, plotting the points which arise from $x = -3, -2\frac{1}{2}, -2, \ldots, 2, 2\frac{1}{2}, 3$.

2. By means of a graph, find the greatest value of the function $y = 2x - x^2$.

3. Find the least value of the function $x^2 - 4x$.

3. The determination of maxima and minima. The diagram (Fig. 18) illustrates the graph of a function

$$y = f(x)$$

with the properties that the value of $f(x)$ at P exceeds that at any point NEAR IT, while the value at Q is less than that at any point NEAR IT. The *function* $f(x)$ has a *maximum* at P and a *minimum* at Q.

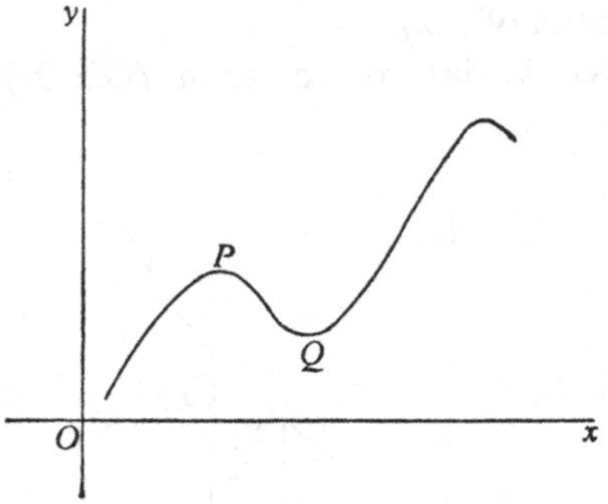

Fig. 18.

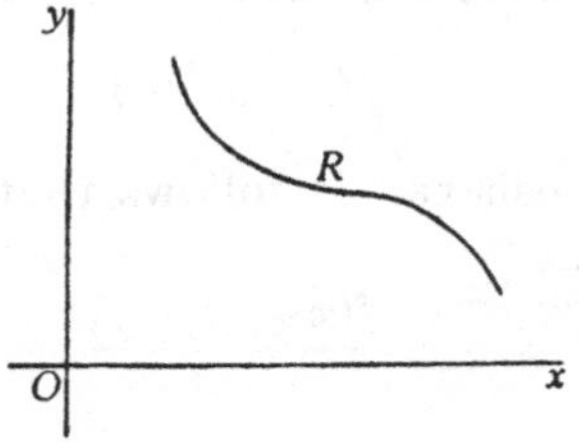

Fig. 19.

It is important to realize that the words *maximum* and *minimum* refer to the parts of the curve near P and Q respectively. Clearly the values at P and Q need not be the greatest or the least values of $f(x)$ over the curve as a whole.

To find the positions of the maxima and minima, we follow precisely the argument of Illustration 1 (p. 49). The criterion is that

the gradient of the curve is zero

at a maximum or minimum. In other words, *if the function* $f(x)$ *has a maximum or a minimum value where* $x = a$, *then*

$$f'(a) = 0.$$

The converse result is NOT necessarily true At the point R in the diagram (Fig. 19) the gradient is zero, although the function has neither a maximum nor a minimum there.

EXAMPLES III

Find the values of x at which each of the following functions has a maximum or minimum value, and illustrate by sketching the curve $y = f(x)$.

1. x^2+1. 2. x^2-2x. 3. x^3-3x-4. 4. x^4-2x^2+7.

5. $\sin x$. 6. $\sin 2x$. 7. $\cos x$. 8. $\cos \frac{1}{2}x$.

4. Increasing and decreasing functions. We should say that, from its appearance, the function $f(x)$, whose graph is shown in the diagram (Fig. 20), is a *steadily increasing function* of x. We must now consider how this feature is to be interpreted in terms of the differential coefficient of $f(x)$.

If x undergoes a small positive variation δx, then $f(x)$, by definition, must increase, so that

$$f(x+\delta x)-f(x)$$

is positive. It follows that the quotient

$$\frac{f(x+\delta x)-f(x)}{\delta x}$$

is essentially positive, and so

$$\lim_{\delta x\to 0}\frac{f(x+\delta x)-f(x)}{\delta x}$$

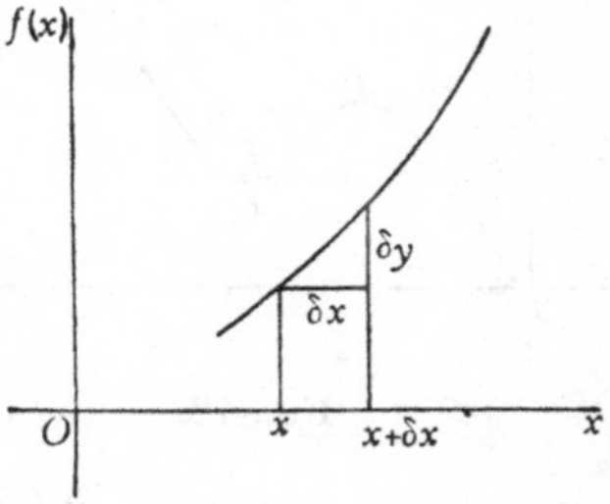

Fig. 20.

is positive. Hence *the differential coefficient of a function is positive in an interval where the function is increasing.*

Similarly, *the differential coefficient of a function is negative in an interval where the function is decreasing.*

The *converses* of these two results are also true.

ILLUSTRATION 2. *To prove that, if x is positive, the value of* $\sin x$ *lies between*

$$x-\frac{x^3}{3!} \quad \textit{and} \quad x-\frac{x^3}{3!}+\frac{x^5}{5!}.$$

Write $$u = \sin x - x + \frac{x^3}{3!}.$$

Then $$u' = \cos x - 1 + \frac{x^2}{2!},$$

$$u'' = -\sin x + x,$$

$$u''' = -\cos x + 1.$$

The differential coefficient of the function u'' is $1-\cos x$, which is necessarily positive; hence u'' is an increasing function of x. But $u''(0) = 0$, and so u'' increases steadily from zero. That is, u'' is positive when x is positive.

The differential coefficient of the function u' is u'', which we have just proved positive; hence u' is an increasing function of x. But $u'(0) = 0$, and so u' increases steadily from zero. That is, u' is positive when x is positive.

Now consider u itself. Its differential coefficient u' has just been proved positive; hence u is an increasing function of x. But $u(0) = 0$, and so u increases steadily from zero. That is, u is positive when x is positive.

Hence $\sin x$ exceeds $x-\frac{x^3}{3!}$ for positive values of x.

In the same way, by considering the function

$$v = x - \frac{x^3}{3!} + \frac{x^5}{5!} - \sin x,$$

we may prove that $\sin x$ is less than $x-\frac{x^3}{3!}+\frac{x^5}{5!}$ for positive values of x.

The result is therefore established.

EXAMPLES IV

1. Prove that, if $x > 0$, then

$$1-\frac{x^2}{2!} < \cos x < 1 - \frac{x^2}{2!} + \frac{x^4}{4!}.$$

2. Prove that, if x lies between 0 and π, then

$$\sin x > x \cos x.$$

3. Prove that, if x lies between 0 and π, then

$$x \sin x + \cos x > 1 + \tfrac{1}{2}x^2 \cos x.$$

4. Find the ranges of values of x for which the following expressions are increasing functions, and illustrate your answers graphically:

(i) $x^2 - x$. (ii) $x^3 + x - 2$. (iii) $x^3 - 3x + 2$.

(iv) $x^2 + 1$. (v) $x^4 - 2x^2$. (vi) $2x^3 - 15x^2 + 24x$.

5. The second differential coefficient and concavity. The diagram (Fig. 21) represents a curve where, in the ordinary sense of the phrase, 'the concavity is upwards'. The curve, as usual, is the graph of a function

$$y = f(x).$$

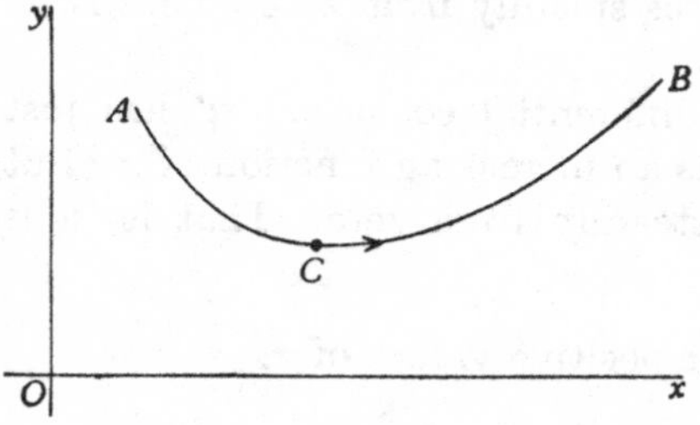

Fig. 21.

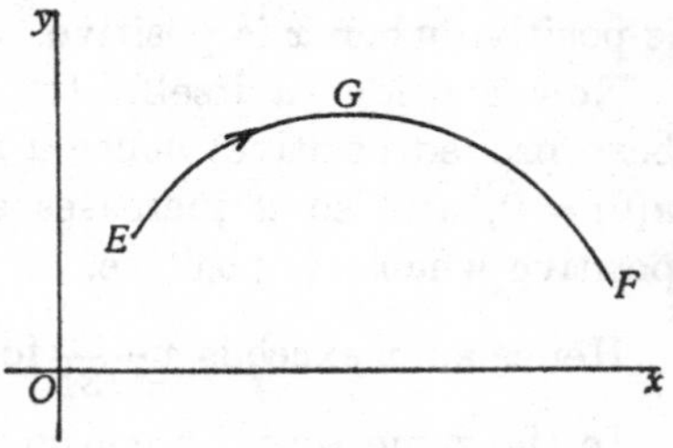

Fig. 22.

As x increases, the corresponding point describes the curve between A and B, moving in the sense indicated by the arrow. The gradient is negative to the left of C and positive to the right, but the important thing is that it INCREASES STEADILY with x. At A the gradient has a fairly high negative value; between A and C it remains negative but decreases in numerical value so that (being negative) it actually increases; at C it has increased to the value zero. After C, the gradient remains positive, and continues to increase.

But the gradient of the function is $f'(x)$, and the condition that $f'(x)$ should increase is that ITS differential coefficient $f''(x)$ should be positive. Hence *the condition for the concavity of the curve* $y = f(x)$ *to be 'upwards' during an interval* $a \leqslant x \leqslant b$ *is that* $f''(x)$ *should be positive in that interval.*

In the same way, the diagram (Fig. 22) represents a curve whose concavity is 'downwards' between E and F. The gradient is positive at E, decreases steadily to zero at G, and thereafter continues to decrease as its value becomes greater and greater with negative sign. Hence *the condition for the concavity of the curve* $y = f(x)$ *to be 'downwards' during an interval* $e \leqslant x \leqslant f$ *is that* $f''(x)$ *should be negative in that interval.*

6. Points of inflexion. The diagrams in § 5 (Figs. 21, 22) illustrated curves in which the concavity was always upwards, or always downwards, in the interval considered. The present diagram (Fig. 23) illustrates a curve

$$y = f(x)$$

in which the concavity is 'downwards' between U and W, but 'upwards' between W and V, changing sense at the point W itself.

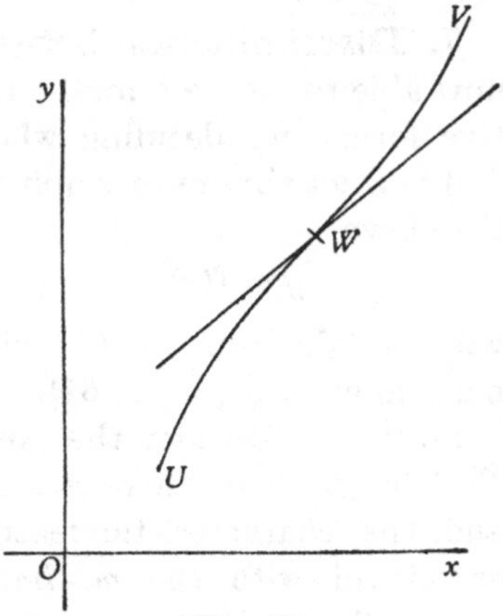

Fig. 23.

The value of $f''(x)$ is negative to the left of W and positive to the right. At W, the value of $f''(x)$ is zero.

It is instructive to trace the progress of the gradient as the curve is traced by a point moving from U, through W, to V. At U the gradient is positive, but it decreases steadily as the point approaches W; after W, however, the gradient begins to increase again, so that it has a minimum value at W. Hence the differential coefficient of the gradient vanishes at W; that is,

$$f''(x) = 0$$

at W.

A point such as W, where the concavity changes sense, is called a *point of inflexion* of the curve.

The condition $f''(x) = 0$ is necessary for a point of inflexion, but not sufficient to ensure it, as the first two examples below illustrate.

EXAMPLES V

1. Sketch the curve $y = \frac{1}{2}x^3$ for values of x between $-2, 2$. Verify that the curve has a point of inflexion at the origin.

2. Sketch the curve $y = \frac{1}{4}x^4$ for values of x between $-2, 2$. Verify that the curve has NOT a point of inflexion at the origin although $\dfrac{d^2y}{dx^2}$ vanishes there.

3. Sketch the curve $y = \frac{1}{5}(x + x^3)$ for values of x between $-3, 3$.

4. Sketch the curve $y = \frac{1}{8}(4x - x^3)$ for values of x between $-4, 4$.

7. Discrimination between maxima and minima. We are now able to devise a method, which can be stated in two alternative forms, for deciding whether a turning value of a function $f(x)$ is a maximum or a minimum. The diagram (Fig. 24) shows the curve

$$y = f(x)$$

with a minimum at P and a maximum at Q (see p. 51).

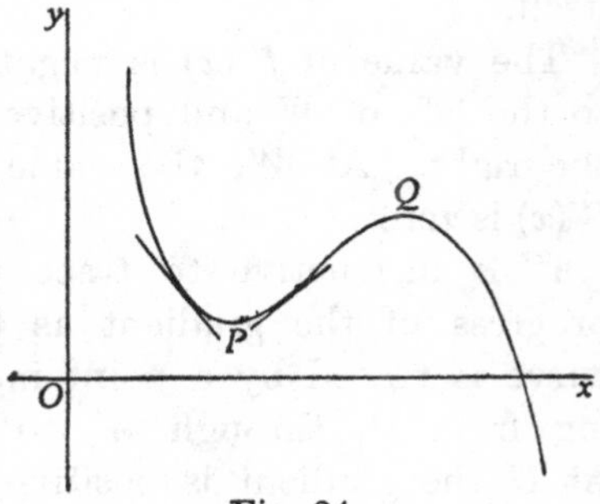

Fig. 24.

FORM I. Consider the point P. The gradient there is zero, and the characteristic feature associated with the *minimum* is that *the gradient passes from negative, through zero, to positive* as x passes, in the increasing sense, through its value at P. Thus:

If $f'(x)$ is negative for values of x just less than x_1, zero at x_1 itself, and positive for values of x just greater than x_1, then $f(x)$ has a minimum at $x = x_1$.

Similarly, *if $f'(x)$ is positive for values of x just less than x_1, zero at x_1 itself, and negative for values of x just greater than x_1, then $f(x)$ has a maximum at $x = x_1$.*

Form II. The first form just given may be reworded to state that the gradient of the curve is an increasing function of x at a minimum and a decreasing function at a maximum. *Hence if* $f''(x)$ *is positive at a turning point, the function has a minimum there; and, if* $f''(x)$ *is negative at a turning point, the function has a maximum there.*

Note. It is often better to use the first form rather than the second, especially if (as often happens) the second differential coefficient is awkward to manipulate.

Illustration 3. *The potential energy of a uniform rod.*

Suppose that OA (Fig. 25) is a uniform rod, of length $2a$ and weight W, suspended from one end O. If it makes an angle θ with the downward vertical, the potential energy V is defined to be the function of θ given by the relation

$$V = \text{const.} - Wa\cos\theta.$$

Hence
$$\frac{dV}{d\theta} = Wa\sin\theta,$$

$$\frac{d^2V}{d\theta^2} = Wa\cos\theta.$$

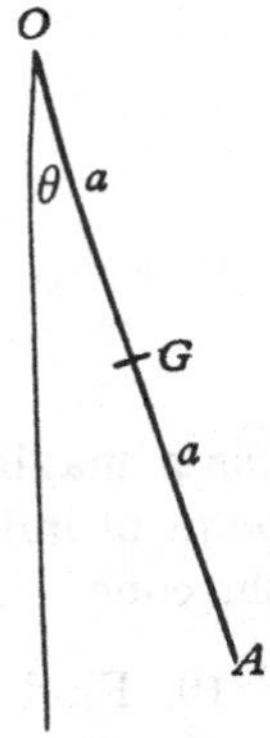

Fig. 25.

The significantly distinct values of θ which give rise to the turning values of the potential energy are derived from the equation

$$Wa\sin\theta = 0,$$

so that
$$\theta = 0 \quad \text{or} \quad \pi.$$

To distinguish between the cases $\theta = 0, \pi$:

Form I.

(i) $\theta = 0$ (*the rod hanging down*).

When θ is just less than zero, $\frac{dV}{d\theta}$ is negative; and, when θ is just greater than zero, $\frac{dV}{d\theta}$ is positive. Hence the potential energy is a minimum.

(ii) $\theta = \pi$ (*the rod 'standing up'*).

When θ is just less than π, $\sin\theta$ is positive, so that $\frac{dV}{d\theta}$ is positive; and, when θ is just greater than π, $\sin\theta$ is negative, so that $\frac{dV}{d\theta}$ is negative. Hence the potential energy is a maximum.

FORM II.

(i) $\theta = 0$.

When $\theta = 0$, $\dfrac{d^2V}{d\theta^2} = Wa$, so that V is a minimum.

(ii) $\theta = \pi$.

When $\theta = \pi$, $\dfrac{d^2V}{d\theta^2} = -Wa$, so that V is a maximum.

EXAMPLES VI

The positions of the maxima and minima of the following functions were found in Examples III (p. 52). Use both forms of test to distinguish between maxima and minima.

1. x^2+1.
2. x^2-2x.
3. x^3-3x-4.
4. x^4-2x^2+7.
5. $\sin x$.
6. $\sin 2x$.
7. $\cos x$.
8. $\cos \frac{1}{2}x$.

9. Prove that the curve

$$y = x^3-3x^2-9x$$

has a maximum where $x=-1$, a minimum where $x=3$, and a point of inflexion where $x=1$. Sketch the curve, and check that the concavity in your sketch agrees with the results of § 5 (p. 54).

10. Find the results analogous to those of Ex. 9 for the curves

(i) $y = x^3-12x$,

(ii) $y = 2x^3+3x^2$.

11. Find the maxima and minima of the function

$$\sin 3x + 3\sin x,$$

and distinguish between them.

12. For the function $y = 2x^3-9x^2+12x$,

(i) find the maxima and minima, and distinguish between them,

(ii) find the values of x for which y is (*a*) an increasing function, (*b*) a decreasing function,

(iii) sketch the curve.

8. The sketching of simple curves. The ability to sketch quickly the principal features of a curve

$$y = f(x)$$

is important for many purposes. The following properties are important:

(i) passage through characteristic points;

(ii) symmetry;

(iii) the sign of $f(x)$ for various ranges of values of x;

(iv) the gradient;

(v) maximum or minimum values of $f(x)$;

(vi) the concavity.

Common sense and experience must decide which of these properties seem likely to be most helpful for a particular curve. The illustration which follows is typical.

ILLUSTRATION 4. *To sketch the curve*

$$y = x^3 - 6x^2 + 9x - 1.$$

By differentiation,

$$y' = 3x^2 - 12x + 9 = 3(x-1)(x-3),$$

$$y'' = 6x - 12 = 6(x-2).$$

The critical features may be expected to be:

(*a*) where $y = 0$, but it is not easy to find such points for this particular curve;

(*b*) where $y' = 0$, that is, where $x = 1, 3$;

(*c*) where $y'' = 0$, that is, where $x = 2$.

We may therefore construct a table as follows:

x	$x<1$	1	$1<x<2$	2	$2<x<3$	3	$x>3$
y	?	3	?	1	?	-1	?
y'	$+$	0	$-$	-3	$-$	0	$+$
y''	$-$	-6	$-$	0	$+$	$+6$	$+$

The concavity is 'downwards' for $x < 2$ and 'upwards' for $x > 2$; there is a point of inflexion at $(2, 1)$, with gradient -3.

There is a maximum at $(1, 3)$ and a minimum at $(3, -1)$.

These features are all gathered together in the diagram (Fig. 26).

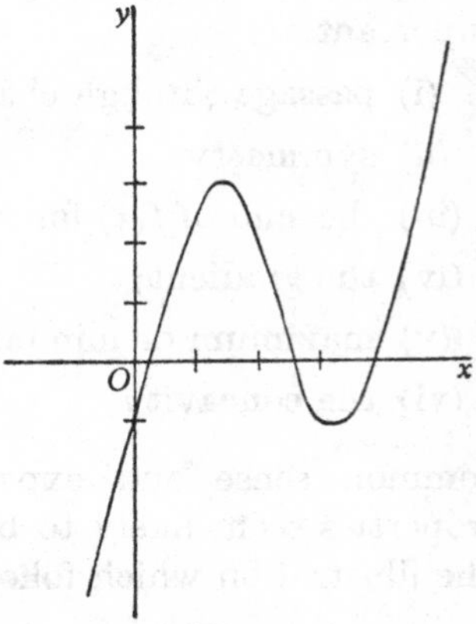

Fig. 26.

EXAMPLES VII

Sketch the curves given by the following equations:

1. $y = x^2$.
2. $y = x^3$.
3. $y = x^2 - 3x + 2$.
4. $y = (x-1)(x-2)(x-3)$.
5. $y = x^3 - 3x$.
6. $y = x^4 - 2x^2$.

See also Examples V (p. 56).

9. Rolle's theorem. To verify that, *if $f(x)$ is a function of x continuous between two points $x = a, x = b$, at each of which it vanishes, the curve $y = f(x)$ having a tangent at all points between a, b, then there exists at least one value of x between a, b at which $f'(x)$ vanishes.*

The diagram (Fig. 27) is typical of the form which the graph $y = f(x)$ must take, cutting the x-axis at the points A, B corresponding to $x = a, b$.

We propose to regard it as obvious from the diagram that there exists between A, B at least one point where the tangent is parallel to the x-axis. For a rigorous proof, the reader should consult a more advanced text-book.

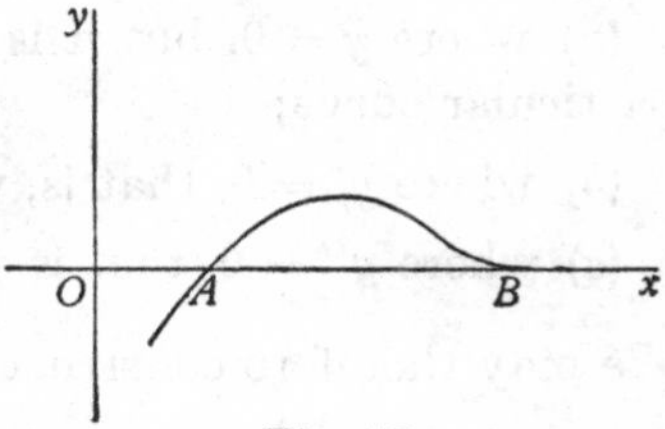

Fig. 27.

EXAMPLES VIII

Verify Rolle's theorem by actual calculation of $f'(x)$, and also by drawing a graph, for each of the following functions:

1. $f(x) \equiv x^2 - 3x + 2$; $a = 1, b = 2$.

2. $f(x) \equiv (x-1)(x-2)(x-3)$; $a = 1, b = 2$.

3. $f(x) \equiv x^2(x-1)$; $a = 0, b = 1$.

4. $f(x) \equiv x(x-1)^2$; $a = 0, b = 1$.

5. $f(x) \equiv \sin x$; $a = \pi, b = 2\pi$.

6. $f(x) \equiv \cos^2 x$; $a = \frac{1}{2}\pi, b = \frac{3}{2}\pi$.

10. The mean value theorem. To verify that, *if $f(x)$ is a function of x (continuous and) having a differential coefficient* at each point between $x = a, x = b$, then there exists a value ξ of x between a, b with the property that*

$$f(b) = f(a) + (b-a)f'(\xi).$$

In the diagram (Fig. 28), let A, B be the points of the curve $y = f(x)$ for which $x = a, b$, and draw AP perpendicular to the ordinate through B. Then

$$\frac{f(b)-f(a)}{b-a} = \frac{PB}{AP} = \tan PAB.$$

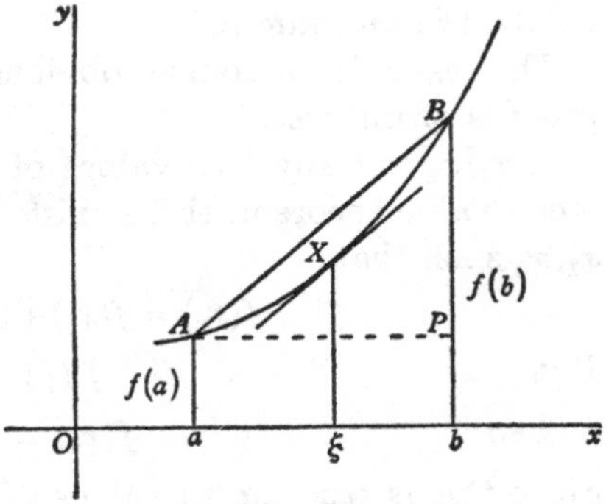

Fig. 28.

But the graph shows that there must be at least one point X between A, B where the tangent is parallel to AB. If this point is given by $x = \xi$, then

$$f'(\xi) = \tan PAB,$$

so that

$$\frac{f(b)-f(a)}{b-a} = f'(\xi).$$

This is equivalent to the required result.

ALITER. The following proof is instructive, and worthy of close study. It is the basis of the proof of a generalization of the mean value theorem to be given later (Vol. II, p. 44).

* Having a differential coefficient ensures that the function is continuous. It should be noted, however, that the converse need not be true—a function may be continuous but not have a differential coefficient. For example, if $y = +x$ when x is positive and $-x$ when x is negative, then (i) y is continuous at $x = 0$; (ii) $y' = +1$ when x is positive; (iii) $y' = -1$ when x is negative; but (iv) y' does not exist when x is zero.

Define a function $u(x)$ as follows:

$$u(x) \equiv f(b) - f(x) - \left\{\frac{f(b)-f(a)}{b-a}\right\}(b-x).$$

Since $f(x)$ is continuous between a, b, so also is $u(x)$. Further, it is easy to see that $u(a) = 0, u(b) = 0$, so that $u(x)$ satisfies the conditions of Rolle's theorem (p. 60). Hence there exists a value ξ of x between a, b such that

$$u'(\xi) = 0.$$

Now
$$u'(x) = -f'(x) + \left\{\frac{f(b)-f(a)}{b-a}\right\},$$

and so
$$f'(\xi) = \frac{f(b)-f(a)}{b-a},$$

as required.

ILLUSTRATION 5. *To prove that, if $f(x)$ is a continuous function of x such that $f'(x) = 0$ throughout a given interval, then $f(x)$ is constant in that interval.*

The result is, of course, obvious graphically, but the following proof is of interest.

If x_1, x_2 are any two values of x in the interval, then, by the mean value theorem, there exists at least one value of ξ between x_1, x_2 such that

$$f(x_2) = f(x_1) + (x_2 - x_1) f'(\xi).$$

But
$$f'(\xi) = 0,$$

and so
$$f(x_2) = f(x_1).$$

Since this is true for all values of x_1, x_2 in the interval, $f(x)$ must be constant.

EXAMPLES IX

1. Prove that, if $f(x)$ is a continuous function of x such that $f'(x) = 2$ throughout a given interval, then $f(x) \equiv 2x + b$ in that interval, where b is some constant.

2. Verify the mean value theorem by calculating ξ for each of the following functions, and illustrate your results graphically.

(i) $f(x) \equiv x^2$; $a = 0, b = 2$.

(ii) $f(x) \equiv x^2 + x$; $a = 1, b = 3$.

(iii) $f(x) \equiv x^3 + 3x$; $a = -1, b = 2$.

11. The real roots of the equations $f(x) = 0, f'(x) = 0$. Suppose that we have to solve an equation $f(x) = 0$, and that the diagram (Fig. 29) represents the curve

$$y = f(x).$$

We restrict ourselves for the moment to the case when the roots are ALL DISTINCT. This restriction, the purpose of which is to ensure that $f'(x)$ and $f(x)$ are not both zero together, is essential for the work which follows. It is assumed, too, that the curve is continuous, with a continuously turning tangent.

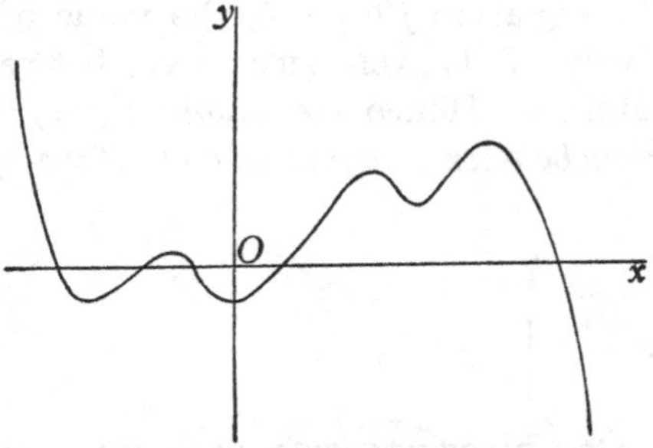

Fig. 29.

Our aim is to discuss, without the analytical detail required for a full study, those properties which may be asserted with reasonable confidence from a study of the graph.

(i) *The sign of $f(x)$.* If x is imagined to increase from $-\infty$ to $+\infty$, the value of $f(x)$ changes sign every time that x passes through a point where $f(x) = 0$. Hence *the number of (real) roots is equal to the number of changes in the sign of $f(x)$ as x increases from $-\infty$ to $+\infty$.*

If $f(a), f(b)$ have the SAME signs, then *there is an even number (or zero) of roots between a, b.*

If $f(a), f(b)$ have OPPOSITE signs, then *there is an odd number of roots between a, b.*

In particular, if

$$f(x) \equiv a_0x^n + a_1x^{n-1} + \ldots + a_n$$

is a polynomial in x of degree n, and if n is odd, then, for large negative values of x, $f(x)$ has sign opposite to a_0, and for large positive values of x, $f(x)$ has the same sign as a_0. Hence *if $f(x)$ is a polynomial of* ODD *degree, the equation $f(x) = 0$ has at least one real root.*

(ii) *The sign of $f'(x)$.* Suppose, for the sake of explanation, that the sign of $f'(x)$ is positive at a point ξ where $f(x) = 0$. Then $f(x)$ is an increasing function of x at $x = \xi$, so that the curve $y = f(x)$

goes 'above' the axis as x increases from ξ. As x, still increasing, approaches the next root $x = \eta$, the curve 'drops' towards the x-axis, so that $f'(x)$ is negative there. Hence *the differential coefficient $f'(x)$ assumes alternate signs at the successive roots of the equation $f(x) = 0$.*

Applying the result (i) above, we see too that *the equation $f'(x) = 0$ has an odd number of roots (at least one) between consecutive roots of the equation $f(x) = 0$.*

(iii) Since $f'(x)$ has constant sign between successive roots of the equation $f'(x) = 0$, the value of $f(x)$ either rises or, alternatively, falls ALL THE WAY between successive maxima and minima. Hence *the equation $f(x) = 0$ has either one root or no roots between consecutive roots of the equation $f'(x) = 0$.*

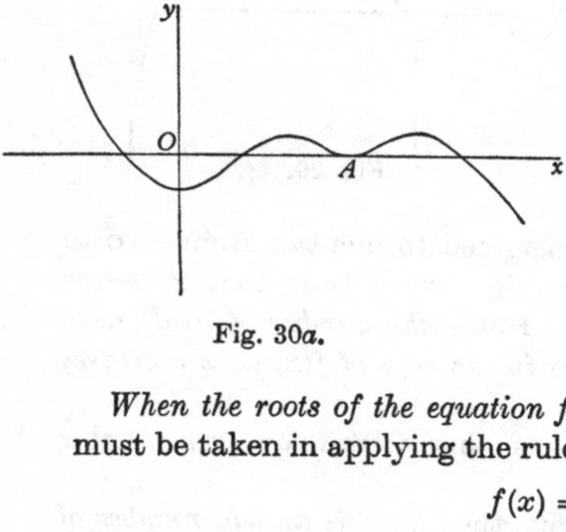

Fig. 30*a*.

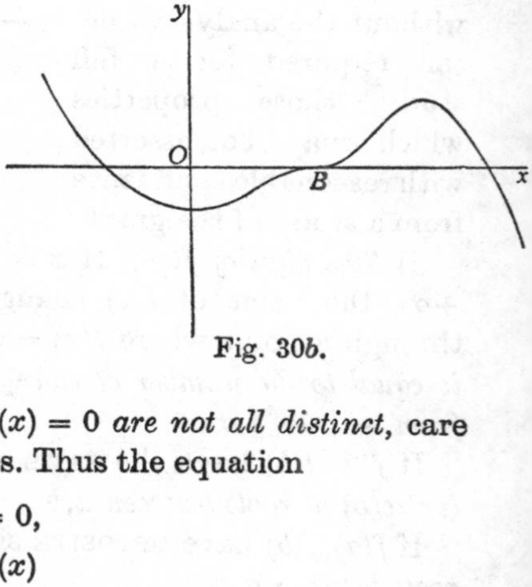

Fig. 30*b*.

When the roots of the equation $f(x) = 0$ are not all distinct, care must be taken in applying the rules. Thus the equation

$$f(x) = 0,$$

whose graph

$$y = f(x)$$

is shown in Fig. 30*a*, has a double root $x = \xi$, corresponding to the point A; and in Fig. 30*b* it has a triple root $x = \eta$ corresponding to the point B. The function $f(x)$ has the same sign for values of x on either side of ξ, but opposite signs for values on either side of η.

It is probably wiser (at any rate, for the present) to treat each case on its merits as it arises rather than to seek a more elaborate set of rules.

ILLUSTRATION 6. *If*

$$P_n(x) \equiv 1 + x + \frac{x^2}{2!} + \ldots + \frac{x^n}{n!},$$

prove that the equation $P_n(x) = 0$ *has no real roots when n is even and exactly one real root when n is odd.*

This is a typical problem. Assume that the result is true for $n = 1, 2, \ldots, N$. We prove that it is then true for $n = N+1$.

(i) *Suppose that N is even.* By the assumption, the equation

$$P_N(x) = 0$$

has no real roots.

Consider the equation

$$P_{N+1}(x) = 0.$$

By direct differentiation, we have the relation

$$P'_{N+1}(x) \equiv P_N(x),$$

so that $P'_{N+1}(x)$ is never zero. Hence the polynomial $P_{N+1}(x)$ either increases steadily or decreases steadily. But, since $N+1$ is odd, $P_{N+1}(x)$ is negative for large negative values of x and positive for large positive values. (Indeed, $P_{N+1}(x)$ is obviously positive and increasing for positive x.) Hence, the graph $y = P_{N+1}(x)$ crosses the x-axis once, and once only, and the equation $P_{N+1}(x) = 0$ has precisely one root.

(ii) *Suppose that N is odd.* By the assumption, the equation

$$P_N(x) = 0$$

has exactly one real root, say ξ, which cannot be zero.

Consider the equation

$$P_{N+1}(x) = 0,$$

where, as before, $$P'_{N+1}(x) \equiv P_N(x),$$

so that $$P'_{N+1}(\xi) = 0.$$

The curve $y = P_{N+1}(x)$ therefore has a turning value at $x = \xi$; and, as this is the *only* turning value, while $P_{N+1}(x)$ is large and positive ($N+1$ being even) for large positive or negative x, the turning point is actually a minimum and gives the least value attained by $P_{N+1}(x)$. Moreover,

$$\begin{aligned} P_{N+1}(\xi) &\equiv \left(1 + \xi + \frac{\xi^2}{2!} + \ldots + \frac{\xi^N}{N!}\right) + \frac{\xi^{N+1}}{(N+1)!} \\ &= \frac{\xi^{N+1}}{(N+1)!} \end{aligned}$$

since $P_N(\xi)$ is zero. Hence, as $N+1$ is even, and ξ is not zero, $P_{N+1}(\xi)$ is positive. Since the least value of $P_{N+1}(x)$ is positive, $P_{N+1}(x)$ can never be zero.

The result is therefore true for $N+1$ whether N is even or odd. But it is clearly true for $n = 1$. Hence it is true for $n = 2, 3, 4, 5, \ldots$, and so generally.

EXAMPLES X

1. Prove that, if $f(x) \equiv x^2(1-x)^2$, then the roots of the equation $f''(x) = 0$ are distinct, and lie between 0 and 1.

Prove the corresponding result for the equation $g'''(x) = 0$, where $g(x) \equiv x^3(1-x)^3$.

2. Indicate in a single diagram the relative positions of the roots of the equations

$$f_1 \equiv x = 0, \quad f_2 \equiv x - \frac{x^3}{3!} = 0, \quad f_3 \equiv x - \frac{x^3}{3!} + \frac{x^5}{5!} = 0,$$

$$g_1 \equiv 1 - \frac{x^2}{2!} = 0, \quad g_2 \equiv 1 - \frac{x^2}{2!} + \frac{x^4}{4!} = 0.$$

12.* Mean value theorems for two functions.

(i) CAUCHY'S MEAN VALUE THEOREM. To prove that, *if the functions $f(x)$, $g(x)$ have differential coefficients which do not vanish simultaneously in the interval a, b, and if $g(a)$ is not equal to $g(b)$, then there is a number ξ between a, b for which*

$$\frac{f(b)-f(a)}{g(b)-g(a)} = \frac{f'(\xi)}{g'(\xi)}.$$

Introduce a function $F(x)$ defined by the relation

$$F(x) \equiv f(x) + Ag(x) + B,$$

where A, B are constants chosen so that

$$F(a) \equiv f(a) + Ag(a) + B = 0,$$

$$F(b) \equiv f(b) + Ag(b) + B = 0.$$

These two equations can be solved for A, B since $g(a)$, $g(b)$ are not equal.

By Rolle's theorem, there is a number ξ between a, b such that $F'(\xi) = 0$. That is,

$$f'(\xi) + Ag'(\xi) = 0.$$

* This paragraph may be postponed, if desired.

Moreover, $g'(\xi)$ cannot be zero; if it were, this equation would make $f'(\xi)$ zero also, contrary to the first hypothesis. Hence we may divide by $g'(\xi)$, giving

$$\frac{f'(\xi)}{g'(\xi)} = -A.$$

But, by direct subtraction,

$$F(b) - F(a) = \{f(b) - f(a)\} + A\{g(b) - g(a)\},$$

so that, since $F(b) = F(a) = 0$ and $g(b) - g(a)$ is not zero,

$$\frac{f(b) - f(a)}{g(b) - g(a)} = -A.$$

Hence
$$\frac{f(b) - f(a)}{g(b) - g(a)} = \frac{f'(\xi)}{g'(\xi)}.$$

(ii) To prove that, *if the functions* $f(x), g(x)$ *have second differential coefficients such that* $f''(x), g''(x)$ *do not vanish simultaneously in the interval* $a-h, a+h$, *and if* $g(a+h) + g(a-h) - 2g(a)$ *is not zero, then there is a number* ξ *between* $a-h, a+h$ *for which*

$$\frac{f(a+h) + f(a-h) - 2f(a)}{g(a+h) + g(a-h) - 2g(a)} = \frac{f''(\xi)}{g''(\xi)}.$$

Introduce a function $F(x)$ defined by the relation

$$F(x) \equiv f(x) + Ag(x) + Bx + C,$$

where A, B, C are constants chosen so that

$$\begin{aligned} F(a+h) &\equiv f(a+h) + Ag(a+h) + B(a+h) + C = 0, \\ F(a-h) &\equiv f(a-h) + Ag(a-h) + B(a-h) + C = 0, \\ F(a) &\equiv f(a) + Ag(a) + Ba + C = 0. \end{aligned}$$

By Rolle's theorem, since $F(a) = F(a+h) = 0$, there is a number ξ_1 between $a, a+h$ such that $F'(\xi_1) = 0$; and, since

$$F(a-h) = F(a) = 0,$$

there is a number ξ_2 between $a-h, a$ such that $F'(\xi_2) = 0$.

Now apply Rolle's theorem to the function $F'(x)$, which vanishes when $x = \xi_1, \xi_2$. There is a number ξ between ξ_1, ξ_2, *and therefore between* $a-h, a+h$, such that $F''(\xi) = 0$. That is

$$f''(\xi) + Ag''(\xi) = 0.$$

But $f''(\xi), g''(\xi)$ are not both zero, by hypothesis, and so $g''(\xi)$ cannot vanish. Hence

$$\frac{f''(\xi)}{g''(\xi)} = -A.$$

Moreover, from the three equations for A, B, C, we have

$$\{f(a+h)+f(a-h)-2f(a)\}+A\{g(a+h)+g(a-h)-2g(a)\} = 0,$$

so that, since $g(a+h)+g(a-h)-2g(a)$ is not zero,

$$\frac{f(a+h)+f(a-h)-2f(a)}{g(a+h)+g(a-h)-2g(a)} = -A$$

$$= \frac{f''(\xi)}{g''(\xi)}.$$

13. Application to certain limits. To prove that, *if $f(x), g(x)$ are (continuous) functions such that $f(a) = g(a) = 0$, and if $f'(a), g'(a)$ both exist ($g'(a) \neq 0$), then*

$$\lim_{x\to a} \frac{f(x)}{g(x)} = \frac{f'(a)}{g'(a)}.$$

Since $f(a) = g(a) = 0$, we have the relation

$$\frac{f(x)}{g(x)} = \frac{\dfrac{f(x)-f(a)}{x-a}}{\dfrac{g(x)-g(a)}{x-a}} \quad (x \neq a).$$

Now let $x \to a$. The right-hand side tends to the ratio (which, by hypothesis, exists) of the two differential coefficients at $x = a$, so that

$$\lim_{x\to a} \frac{f(x)}{g(x)} = \frac{f'(a)}{g'(a)}.$$

EXTENSION.* More generally, *if $f(x), g(x)$ are (continuous) functions such that $f(a) = g(a) = 0$, and if $f'(x)/g'(x)$ tends to the limit l as x tends to a, then*

$$\lim_{x\to a} \frac{f(x)}{g(x)} = l.$$

To prove this, we use Cauchy's mean value theorem (p. 66), that

$$\frac{f(x)-f(a)}{g(x)-g(a)} = \frac{f'(\xi)}{g'(\xi)}$$

* This extension may be postponed, if desired.

for a number ξ between x, a. As x tends to a, the number ξ must tend to a also, and so

$$\lim_{x \to a} \frac{f(x)}{g(x)} = \lim_{x \to a} \frac{f(x) - f(a)}{g(x) - g(a)}$$

$$= \lim_{x \to a} \frac{f'(\xi)}{g'(\xi)}$$

$$= \lim_{x \to a} \frac{f'(x)}{g'(x)}.$$

NOTE. This extension (which the reader who has not yet studied Cauchy's mean value theorem may accept without proof) leads to a 'continuation' process for evaluating the limit $f(x)/g(x)$, assuming existence and continuity where necessary:

(i) If $f(a) = g(a) = 0$, but $g'(a) \neq 0$, then

$$\lim_{x \to a} \frac{f(x)}{g(x)} = \frac{f'(a)}{g'(a)}.$$

(ii) If also $f'(a) = g'(a) = 0$, but $g''(a) \neq 0$, then

$$\lim_{x \to a} \frac{f'(x)}{g'(x)} = \lim_{x \to a} \frac{f''(x)}{g''(x)},$$

so that
$$\lim_{x \to a} \frac{f(x)}{g(x)} = \frac{f''(a)}{g''(a)},$$

and so on.

ILLUSTRATION 7. *To evaluate the limit*

$$\lim_{x \to 0} \frac{1 - \cos x}{x^2}.$$

If
$$f(x) \equiv 1 - \cos x, \quad g(x) = x^2,$$

then $f(0) = g(0) = 0$. Also

$$f'(x) = \sin x, \quad g'(x) = 2x,$$

so that $f'(0) = g'(0) = 0$. We therefore proceed to the next stage

$$f''(x) = \cos x, \quad g''(x) = 2,$$

so that
$$f''(0) = 1, \quad g''(0) = 2.$$

Hence
$$\lim_{x \to 0} \frac{1 - \cos x}{x^2} = \frac{1}{2}.$$

EXAMPLES XI

Use the method of § 13 to evaluate the following limits:

1. $\lim_{x\to 1} \dfrac{1-x^3}{1-x^2}$.

2. $\lim_{x\to 1} \dfrac{(1-x)^3}{1-x^3}$.

3. $\lim_{x\to 2} \dfrac{\sin \pi x}{x-2}$.

4. $\lim_{x\to 0} \dfrac{\sin x}{x}$.

5. $\lim_{x\to 0} \dfrac{x-\sin x}{x^3}$.

6. $\lim_{x\to 0} \dfrac{\sin^2 x}{x^2}$.

REVISION EXAMPLES I

'Alternative Ordinary' Level

1. Differentiate the following functions with respect to x:

(i) $\cos x + x \sin x$, (ii) $(3x-1)(x-3)$, (iii) $1/(1+x^2)$.

2. Differentiate with respect to x:

(i) $x^2\sqrt{(1-x)}$, (ii) $\sin^2 x \cos^2 x$, (iii) $\dfrac{x-1}{x^2+1}$.

3. Differentiate with respect to x:

(i) $\dfrac{(x^2-1)(x^2-2)}{x^2}$, (ii) $\tan^2 x$.

4. Prove from first principles that the differential coefficient of $1/x^2$ with respect to x is $-2/x^3$.

Differentiate with respect to x:

(i) $\left(x+\dfrac{1}{x}\right)^2$, (ii) $\dfrac{\sin x}{1+\sin x}$, (iii) $\sqrt{(a^2-x^2)}$.

5. Differentiate with respect to x:

(i) $x^3(1+x)^2$, (ii) $\sin^2 2x$, (iii) $\sqrt{(1-2x^2)}$.

6. Differentiate the following with respect to x:

(i) $x^2(3-2x)$, (ii) $\sin 2x \cos x$, (iii) $1/\sqrt{(1-x^2)}$.

7. Find the equation of the tangent to $y = x^2$ at the point $(1, 1)$ and of the tangent to $y = \frac{1}{6}x^3$ at the point $(2, \frac{4}{3})$.

Show that these tangents are parallel and that the distance between them is $\frac{1}{3}\sqrt{5}$.

8. The curve $$y = x(x-1)(x-2)$$
cuts the x-axis at the points $O(0,0), A(1,0), B(2,0)$. Find the equations of the tangents at O, A, B.

Find the coordinates of the point at which the tangent at A cuts the tangent at B.

9. Find the x-coordinates of the points on the curve
$$y = (x+1)(x-2)^2$$
at which the gradient is zero, and test whether y has a maximum or minimum value at each of the points that you have found.

Also find the x-coordinate of the point at which the tangent is parallel to the tangent at the point $(3, 4)$.

Draw a rough sketch of the curve.

10. Find the equation of the tangent to the curve
$$y = x^3 + \tfrac{1}{2}x^2 + 1$$
at the point $(-1, \tfrac{1}{2})$.

Find the coordinates of another point on the curve where the tangent is parallel to that at the point $(-1, \tfrac{1}{2})$.

11. Find the coordinates of the points of intersection of the line $3y = x$ with the curve
$$y = x(1-x^2).$$
If these points are in order P, O, Q, prove that the tangents to the curve at P, Q are parallel, and that the tangent at O is perpendicular to them.

12. Find the equation of the tangent to the curve
$$y = x^3 - 9x^2 + 20x - 8$$
at the point $(1, 4)$.

At what points of the curve is the tangent parallel to the line $4x + y = 3$?

13. The point (h, k) lies on the curve $y = 2x^2 + 18$. Find the gradient at this point and the equation of the tangent there.

Hence find the equations of the two tangents to the curve which pass through the origin.

14. Prove that the equation of the tangent to the curve

$$y = 2x^3 + 2x^2 - 8x + 7$$

at the point (1, 3) is $2x - y + 1 = 0$.

Find also the abscissae (x-coordinates) of the points on the curve at which the tangents are perpendicular to the first tangent.

15. A curve whose equation is

$$12y = ax^3 + bx^2 + cx + d$$

has the following properties:

(i) it passes through the origin, and the tangent there makes an angle of 45° with the axis OX;

(ii) it is parallel to the axis OX when $x = 1$ and when $x = 2$; determine a, b, c, d and sketch the curve.

16. Find the maximum and minimum values of the expression

$$x^4 + 4x^3 - 2x^2 - 12x + 2,$$

distinguishing maxima from minima.

17. Find the maximum and minimum values of the function

$$\tfrac{1}{8}(x-2)^2(x+4),$$

carefully distinguishing between them.

Use these results to sketch the graph of the function, and find the equation of the tangent at the point (0, 2) on the graph.

18. Find the points on the curve

$$y = 2x^3 - 3x^2 - 12x + 20$$

at which y has a maximum or minimum value.

Use your results to make a rough sketch of the curve.

19. Find the points (x, y) on the curve whose equation is

$$y = x^3 - 6x^2 + 9x + 2$$

at which y is a maximum or minimum.

Use your results to draw a rough sketch of this curve.

20. Find the gradient of the curve

$$y = 2x^3 - 5x^2 - 4x + 12$$

at the point $(2, 0)$ and determine whether y has a maximum or a minimum value at this point. Illustrate your answer by a rough sketch showing the slope of the curve in the neighbourhood of the point.

21. Find the coordinates of the points on the curve

$$y = x^3 - 6x^2 + 9x - 2$$

at which the tangents are parallel to the x-axis.

Show that the slopes of the tangents at points whose abscissae (x-coordinates) are less than 1 are all positive, and that at points whose abscissae lie between 1 and 3 the slopes are negative.

Sketch the curve, and state what you infer as to the nature of the roots of the equation

$$x^3 - 6x^2 + 9x - 2 = 0.$$

22. A particle is moving along a straight line in such a manner that its distance from a point O on the line at time t sec. is given by $x = pt^2 + qt^3$, where p and q are constants. Find the velocity and acceleration of the particle at time t in terms of p, q, t.

Find also the values of p and q for which the maximum velocity is 48 ft./sec., this velocity being attained when $t = 4$.

23. OX and OY are two perpendicular straight roads, and A is a fixed point on OX, distant a ft. from O. A motor-cyclist P is travelling along OY at a constant speed v ft./sec., and at time t the angle OAP is θ radians. Prove that the rate of increase of the distance of P from A is $v \sin \theta$ ft./sec. and that the rate of increase of the angle θ is $(v/a)\cos^2\theta$ radians/sec.

24. A particle moves along the x-axis in such a way that its distance x ft. from the origin after t sec. is given by the formula $x = 27t - 2t^2$. What are its velocity and acceleration after $6\frac{3}{4}$ sec?

How long does it take for the velocity to be reduced from 15 ft. per sec. to 9 ft. per sec., and how far does the particle travel meanwhile?

25. A point moves along a straight line so that, at the end of t sec., its distance from a fixed point on the line is $t^3 - 2t^2 + t$ ft. Find the velocity and acceleration at the end of 3 sec.

26. A particle moves along a straight line so that its distance at time t from a given point O of the line is x, where $x = t \sin t + \cos t$.

Find its velocity and acceleration at time t.

Prove that the particle is at rest at the times

$$t = 0, \tfrac{1}{2}\pi, \tfrac{3}{2}\pi, \tfrac{5}{2}\pi, \tfrac{7}{2}\pi, \ldots.$$

27. The angle C of the triangle ABC is always right. If the sum of CA, CB is 6 in., find the maximum area of the triangle.

If, on the other hand, the hypotenuse AB is kept equal to 4 in. and the sides CA, CB allowed to vary, find the maximum area of the triangle.

28. A man wishes to fence in a rectangular closure of area 128 sq. ft. One side of the enclosure is formed by part of a brick wall, already in position. What is the least possible length of fencing required for the other three sides? (Prove that your result gives a *minimum*.)

29. B is a point a miles due north of A, while C is $3a$ miles due east of B; P is a variable point on BC at a distance x miles from B. A man walks straight from A to P at 4 miles per hour and then straight from P to C at 5 miles per hour. Prove that the time for the whole journey is, in hours,

$$\tfrac{1}{4}\sqrt{(a^2+x^2)} + \tfrac{1}{5}(3a-x).$$

Find what the value of x must be for the time taken on the whole journey to be a minimum, and also find this minimum time in hours.

30. A statue 12 ft. high stands on a pillar 14 ft. high. A man, whose eye is 5 ft. above the ground, stands at a distance x ft. from the statue. Prove that the angle θ which the statue subtends at his eye is given by the equation

$$\tan\theta = \frac{12x}{x^2+189}.$$

Find the value of x for which θ is as great as possible, giving your answer correct to one place of decimals.

31. A rectangular tank, open at the top, on a base x ft. by y ft. and of height x ft., is to be constructed of iron sheeting (whose thickness may be neglected) of total area 1350 sq. ft., so that the volume of water which it can contain is a maximum. Find this maximum volume.

32. Post Office Regulations restrict parcels to a maximum length of 3 ft. 6 in. and a maximum girth of 6 ft. Find the maximum permissible volume of a rectangular parcel.

Find also the length of the longest thin rod which can be packed inside a parcel of maximum permissible volume, giving your answer in feet to three significant figures.

33. An open tank is to be constructed with a horizontal square base and four vertical rectangular sides. It is to have a capacity of 32 cu. ft. Find the least area of sheet metal of which it can be made.

34. A piece of wire of length l is cut into two portions, the length of one being x. Each portion is then bent to form the perimeter of a rectangle whose length is twice its breadth. Find an expression for the sum of the areas of these rectangles.

For what value of x is this area a minimum?

35. Prove that, if the sum of the radii of two circles remains constant, the sum of the areas of the circles is least when the circles are equal.

36. A farmer has a certain length of fencing and uses it all to fence in two square sheep-folds. Prove that the sum of the areas of the two folds is least when their sides are equal.

37. A piece of wire of length l is cut into parts of lengths x and $l-x$. The former is bent into the shape of a square, and the latter into a rectangle of which the base is double the height. Find an expression for the sum of the areas of these two figures.

Prove that the only value of x for which this sum is a maximum or a minimum is $x = \frac{8}{17}l$; and find which it is.

38. The vertex and circumference of the base of a right circular cone lie on the surface of a sphere of radius R. The centre of the sphere lies inside the cone. It is a known result that, if the height of the cone is x, its volume is $\frac{1}{3}\pi x^2(2R-x)$. Prove that, if R is unaltered while x is increased by a small quantity h, the volume of the cone is increased by $\frac{1}{3}\pi x(4R-3x)h$ approximately.

39. A cubical block of metal is heated and expands slightly. If its volume increases by h per cent, show that the length of each edge increases by $\frac{1}{3}h$ per cent approximately.

By what approximate percentage is the surface area increased?

40. A body consists of a cylinder of variable radius x in. and fixed length 10 in. closed at each end by a hemispherical cap of radius x in. The volume of the body increases at the rate of 144 cu. in. per sec. At what rate per sec. is x increasing when $x = 3$?

41. It is known that, if l in. is the length of a pendulum and t sec. the time of one complete swing of the pendulum, l is proportional to t^2. If the length of the pendulum is increased by h per cent, where h is small, show that the time of the swing is increased by $\frac{1}{2}h$ per cent approximately.

A clock loses 30 sec. per day of 24 hr. Should the pendulum be lengthened or shortened to make the clock keep correct time, and by what percentage?

42. The height of a closed cylinder is 3 in., and remains constant. The radius of its base is 2·5 in. and it is increasing at the rate of 0·01 in. per sec. At what rates are (i) the volume, (ii) the total surface of the cylinder increasing?

43. A sphere is expanding so that its surface is increasing at the rate of 0·01 sq. in. per sec. Taking $\pi = \frac{22}{7}$, find the approximate rates of increase of (i) its radius, and (ii) its volume at the instant when its radius is 5 in.

44. The volume of a spherical lump of ice t hours after it has begun to melt is V cu. in., its surface is S sq. in. and its radius is r in. If $\frac{dV}{dt} = -3S$, find $\frac{dr}{dt}$.

What is the rate of decrease of S in square inches per hour when $r = 1$?

45. A solid rectangular block of metal expands by being heated through a certain range of temperature, the percentage increase in its length, breadth and thickness being the same. It is found that the percentage increase in its volume is h, where h is small. Find the approximate percentage increase in the length of the block.

CHAPTER IV

THE IDEA OF INTEGRATION

1. The area 'under' a curve. The diagram (Fig. 31) represents the graph of the function

$$y = f(x)$$

for values of x between a, b. We assume for the moment that the curve rises steadily from its value at a to its value at b, and lies

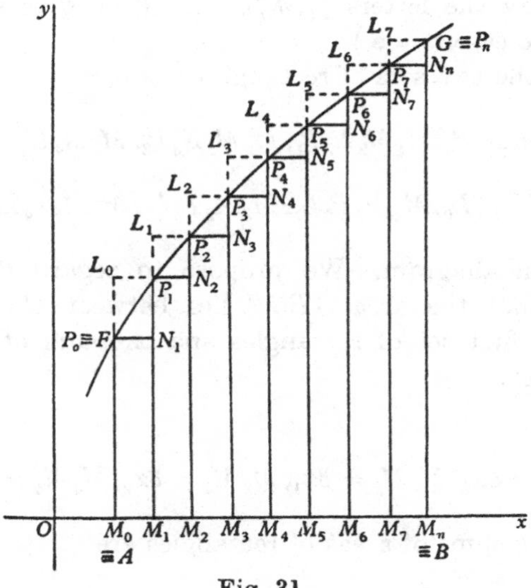

Fig. 31.

entirely above the x-axis* The ordinates AF, BG are drawn for $x = a, b$ respectively.

Our problem is *to find an expression for the area of the figure* $ABGF$.

The basic definition of an area to which we appeal is that, if a rectangle is drawn with sides of lengths p, q, then its area is pq.

* An example will be given later (p. 84), when the reader has acquired more experience, to illustrate the treatment when some of the curve FG lies above, and some below, the x-axis.

We therefore seek to attach rectangles as closely as may be to the figure $ABGF$.

Divide the segment AB of the x-axis into n parts, not necessarily equal in length,

$$AM_1, M_1M_2, M_2M_3, \ldots, M_{n-2}M_{n-1}, M_{n-1}B.$$

For convenience of notation, call A, B the points M_0, M_n; then the intervals are

$$M_0M_1, M_1M_2, M_2M_3, \ldots, M_{n-2}M_{n-1}, M_{n-1}M_n.$$

Through each point $M_0, M_1, M_2, \ldots$ draw the ordinate, meeting the curve in points $P_0, P_1, P_2, \ldots$; let the lengths $M_0P_0, M_1P_1, M_2P_2, \ldots$ be denoted by the letters $y_0, y_1, y_2, \ldots$. (The diagram (Fig. 31) illustrates the case $n = 8$.)

Complete the two sets of rectangles

$$M_0M_1N_1P_0, M_1M_2N_2P_1, M_2M_3N_3P_2, M_3M_4N_4P_3, \ldots$$

and $$M_0M_1P_1L_0, M_1M_2P_2L_1, M_2M_3P_3L_2, M_3M_4P_4L_3, \ldots$$

shown in the diagram. We propose to regard it as obvious intuitively that the area $ABGF$ lies between the sum of the areas of the first set of rectangles and the sum of the areas of the second set.

Now write

$$M_0M_1 = \delta x_0, M_1M_2 = \delta x_1, M_2M_3 = \delta x_2, M_3M_4 = \delta x_3, \ldots.$$

Then the two sums of areas of rectangles are

$$y_0\,\delta x_0 + y_1\,\delta x_1 + y_2\,\delta x_2 + y_3\,\delta x_3 + \ldots$$

and $$y_1\,\delta x_0 + y_2\,\delta x_1 + y_3\,\delta x_2 + y_4\,\delta x_3 + \ldots,$$

or, in more concise notation,

$$\sum_0^{n-1} y_i\,\delta x_i,$$

$$\sum_0^{n-1} y_{i+1}\,\delta x_i.$$

Thus, if Δ is the required area,

$$\sum_0^{n-1} y_i\,\delta x_i < \Delta < \sum_0^{n-1} y_{i+1}\,\delta x_i.$$

We may now suppose that the number n of intervals is taken larger and larger, while the length of each interval becomes smaller and smaller. Then the area of a typical rectangle of the first set, say $M_iM_{i+1}N_{i+1}P_i$ becomes progressively nearer and nearer to the area of the corresponding rectangle $M_iM_{i+1}P_{i+1}L_i$ of the second set. In the limit, as the number of intervals increases indefinitely, their size decreasing indefinitely, the two sums

$$\sum_0^{n-1} y_i\,\delta x_i,\quad \sum_0^{n-1} y_{i+1}\,\delta x_i$$

approach equality, the sum of the areas of the small rectangles 'covering' the arc FG shrinking to zero.

This common limiting value is known as the *area* of $ABGF$.

ILLUSTRATION 1. *To find the area 'under' the curve*

$$y = \tfrac{1}{4}x^2$$

between the ordinates $x = 1, x = 2$.

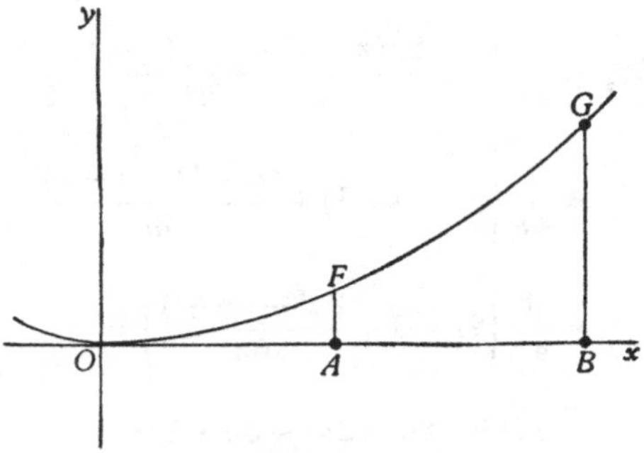

Fig. 32.

Divide the interval between the two points $A(1, 0)$, $B(2, 0)$ into n equal parts (Fig. 32). The points of division occur where

$$x = 1, 1+\frac{1}{n}, 1+\frac{2}{n}, \ldots, 1+\frac{(n-1)}{n}, 2.$$

In the notation of the preceding paragraph,

$$\delta x_0 = \delta x_1 = \delta x_2 = \ldots = \frac{1}{n},$$

$$y_0 = \frac{1}{4},\ y_1 = \frac{1}{4}\left(1+\frac{1}{n}\right)^2,\ y_2 = \frac{1}{4}\left(1+\frac{2}{n}\right)^2,\ y_3 = \frac{1}{4}\left(1+\frac{3}{n}\right)^2, \ldots.$$

Hence
$$\sum_0^{n-1} y_i\,\delta x_i = \sum_0^{n-1} \frac{1}{4}\left(1+\frac{i}{n}\right)^2 \frac{1}{n}$$
$$= \frac{1}{4n}\sum_0^{n-1}\left(1+\frac{2i}{n}+\frac{i^2}{n^2}\right).$$

Now it is proved in text-books on algebra that

$$\sum_0^k 1 = k+1,$$

$$\sum_0^k i = \tfrac{1}{2}k(k+1),$$

$$\sum_0^k i^2 = \tfrac{1}{6}k(k+1)(2k+1),$$

so that

$$\sum_0^{n-1} y_i\,\delta x_i = \frac{1}{4n}\left\{n+\frac{2}{n}.\frac{(n-1)n}{2}+\frac{1}{n^2}\frac{(n-1)n(2n-1)}{6}\right\}$$

$$= \frac{1}{4n}\left\{n+(n-1)+\frac{(n-1)(2n-1)}{6n}\right\}$$

$$= \frac{1}{4n}\left\{2n-1+\frac{2n^2-3n+1}{6n}\right\}$$

$$= \frac{12n^2-6n+2n^2-3n+1}{24n^2}$$

$$= \frac{14n^2-9n+1}{24n^2}$$

$$= \frac{7}{12}-\frac{3}{8n}+\frac{1}{24n^2}.$$

If n increases indefinitely, the number of intervals becomes larger and larger while their magnitude becomes smaller and smaller. In the limit, as $n \to \infty$,

$$\sum_0^{n-1} y_i \delta x_i \to \frac{7}{12}.$$

In the same way,

$$\sum_0^{n-1} y_{i+1} \delta x_i = \sum_0^{n-1} \frac{1}{4}\left(1+\frac{i+1}{n}\right)^2 \frac{1}{n}.$$

It can be proved similarly that

$$\sum_0^{n-1} y_{i+1} \delta x_i \to \frac{7}{12}.$$

Thus the two sums have the same limit, which is the area of the figure $ABGF$. Hence the area is $\frac{7}{12}$ square units.

EXAMPLES I

1. Prove that $$\sum_0^{n-1} y_{i+1} \delta x_i \to \frac{7}{12}.$$

2. Use the method of Illustration 1 (p. 79) to prove that the areas 'under' the curves $y = x$ and $y = x^3$ between $x = 1$ and $x = 2$ are $\frac{3}{2}$ and $\frac{15}{4}$ respectively. [You are given that $\sum_0^k i^3 = \frac{1}{4}k^2(k+1)^2$.]

2. The integral. The ideas which we used in § 1 can be put in a somewhat more general form. Suppose that $f(x)$ is a function of x defined between the values $x = a, b$. As before, we divide the interval into n parts, not necessarily equal, at points where x assumes in turn the values

$$a, x_1, x_2, x_3, \ldots, x_{n-2}, x_{n-1}, b.$$

For convenience of notation, we also write $x_0 = a$, $x_n = b$.

Consider a typical interval (x_i, x_{i+1}), whose length we denote by the symbol δx_i, so that

$$\delta x_i = x_{i+1} - x_i.$$

Throughout that interval, $f(x)$ assumes a succession of values, and these will (for ordinary functions) have a greatest value and a least value. We denote the least value of $f(x)$ in the interval (x_i, x_{i+1}) by the symbol m_i, and the greatest value by M_i.

Now form the two sums

$$s_n = \sum_0^{n-1} m_i \, \delta x_i,$$

$$S_n = \sum_0^{n-1} M_i \, \delta x_i.$$

These correspond to the sums

$$\sum_0^{n-1} y_i \, \delta x_i, \quad \sum_0^{n-1} y_{i+1} \, \delta x_i$$

respectively in § 1 (p. 78); but here we are not assuming that the least and greatest values m_i, M_i in the interval (x_i, x_{i+1}) occur at its end points.

Suppose that the number of intervals is taken larger and larger, while the magnitude of each interval becomes smaller and smaller. The two sums s_n, S_n tend to limits which we call s, S respectively. We assume without proof the theorem that the values s, S are independent of the way in which the subdivisions proceed to their limit.

DEFINITION. *When the two limiting sums s, S are equal, the function $f(x)$ is said to be* INTEGRABLE *between a, b, and their common value is called the integral of $f(x)$ between the limits a, b.*

Thus Illustration 1 (p. 79) tells us that the function $\frac{1}{4}x^2$ is integrable between the limits 1, 2, and that the value of the integral of $\frac{1}{4}x^2$ between the limits 1, 2 is $\frac{7}{12}$.

COROLLARY. If we replace m_i or M_i in the definitions by $f(\xi_i)$, where ξ_i is any point of the interval δx_i, then

$$\sum_0^{n-1} f(\xi_i) \, \delta x_i$$

lies, by definition of m_i, M_i, between the two sums

$$\sum_0^{n-1} m_i \, \delta x_i, \quad \sum_0^{n-1} M_i \, \delta x_i.$$

Hence the sum $$\sum_0^{n-1} f(\xi_i) \, \delta x_i$$

also has as its limit the integral of $f(x)$ between the limits a, b.

3. Notation. We have seen that the integral of a function $f(x)$ between the limits a, b is the common limiting value assumed by the sums

$$\sum_0^{n-1} m_i \,\delta x_i,$$

$$\sum_0^{n-1} M_i \,\delta x_i,$$

or, by the preceding corollary,

$$\sum_0^{n-1} f(\xi_i) \,\delta x_i.$$

To imply that we are to sum over an 'infinite' number of intervals, we replace the *symbol of summation* Σ by a *symbol of integration* $\int$; we also replace the particular x_i by the current variable x, and, instead of the notation δx_i for the very small increment at the point x_i, we write the symbol dx. The result is the notation

$$\int_a^b f(x)\,dx$$

to denote the integral of $f(x)$ between the limits a, b. For example (p. 81),

$$\int_1^2 \frac{1}{4} x^2 \,dx = \frac{7}{12}.$$

COROLLARY (i). It is an immediate consequence of the definition of an integral as summation over sub-intervals that

$$\int_a^c f(x)\,dx + \int_c^b f(x)\,dx = \int_a^b f(x)\,dx.$$

COROLLARY (ii). When $a = b$, it is clear that

$$\int_a^a f(x)\,dx = 0.$$

COROLLARY (iii). We can use Corollaries (i), (ii) to give us an interpretation for the integral

$$\int_b^a f(x)\,dx \quad (a < b),$$

which has not yet been defined. We adopt the meaning, consistent with (i), that

$$\int_b^a f(x)\,dx + \int_a^b f(x)\,dx = \int_b^b f(x)\,dx$$
$$= 0 \quad \text{(by (ii))}.$$

That is to say, $$\int_b^a f(x)\,dx = -\int_a^b f(x)\,dx.$$

Thus *the value of an integral is changed in sign by interchange of the limits.*

4. The sign of the area. The value of an integral, defined as the limit of the sum

$$\sum_0^{n-1} f(x_i)\,\delta x_i$$

may easily be negative in cases where $f(x)$ is not restricted to a positive value.

Suppose, for example, that the function $f(x)$, whose graph

$$y = f(x)$$

is illustrated in the diagram (Fig. 33), is positive in the interval (a, c) and negative in (c, b), being zero at c where the curve crosses the x-axis. If A, B are the values (essentially positive) of the two areas indicated, then

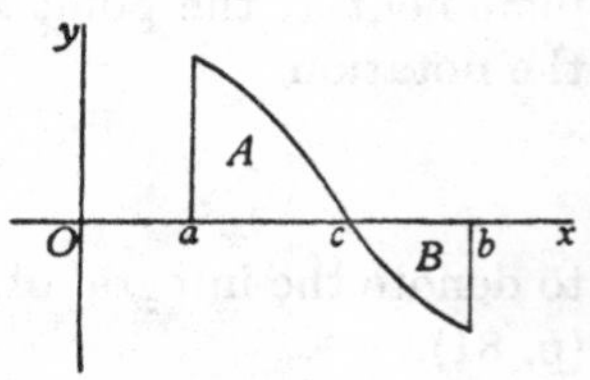

Fig. 33.

$$A = \int_a^c f(x)\,dx$$

as before; but $$B = -\int_c^b f(x)\,dx,$$

since the area is the limiting value of the areas of a number of rectangles of typical length δx_i and 'height' $-f(x_i)$, this being the *numerical value* of the negative number $f(x_i)$. Hence

$$\int_a^b f(x)\,dx = \int_a^c f(x)\,dx + \int_c^b f(x)\,dx$$
$$= A - B,$$

and this may be positive, negative, or zero.

It follows that *it is always wise to sketch a diagram when calculating the area 'under' a curve.*

Further, and more detailed, treatment of areas contained by *closed* curves will be given later (Vol. II, p. 128).

5. Definite and indefinite integrals. Consider the integral

$$\int_a^b f(x)\,dx.$$

Its value depends on each of the limits a, b, so that it is, in fact, a certain function of a, b.

Suppose that the lower limit a has an assigned value, but that the upper limit b is subject to variation. In order to imply this ability to vary, we change the name from b to $\bar{x}$. The value of the integral is then a function of $\bar{x}$, and we may write

$$u(\bar{x}) \equiv \int_a^{\bar{x}} f(x)\,dx.$$

Having done this, it is customary to drop the bar from the symbol $\bar{x}$, thus denoting the upper limit by the letter x itself, and to write

$$u(x) \equiv \int_a^x f(x)\,dx.$$

The symbol x is now doing double duty, as the variable in the function integrated, and as the upper limit of integration. This double use sometimes troubles beginners, and it should be watched carefully. The point is sometimes met by changing the name of the variable in the function integrated to the letter t. We then have

$$u(x) \equiv \int_a^x f(t)\,dt.$$

[It may be useful to point out explicitly that the value of an integral is not affected by changing the name of the variable integrated, the limits being unaltered. Thus

$$\int_a^x f(t)\,dt = \int_a^x f(u)\,du.]$$

We now notice that a change in the *lower* limit a merely affects the value of $u(x)$ to the extent of an additive constant. For let a_1, a_2 be selected as two values of the lower limit. By Corollary (i), p. 83, we have the relation

$$\int_{a_1}^{x} f(x)\,dx = \int_{a_1}^{a_2} f(x)\,dx + \int_{a_2}^{x} f(x)\,dx,$$

and the expression $\int_{a_1}^{a_2} f(x)\,dx$ is a constant, unaffected by a change in the value of the upper limit x in the other integrals.

Finally, it is customary in many problems to leave the lower limit unspecified. We then obtain an integral which might be written

$$\int^{x} f(x)\,dx,$$

but which is in practice written in the simple form

$$\int f(x)\,dx$$

with limits omitted. This is a function of x, determined to within an additive constant whose magnitude is arbitrary because of the unspecified lower limit.

The integral
$$\int_{a}^{b} f(x)\,dx$$

with given limits is called a *definite* integral, and the integral

$$\int f(x)\,dx$$

with unspecified limits is called an *indefinite* integral.

6. The evaluation of an integral. The process of calculating the value of an integral by subdividing an interval, summing, and proceeding to the limit is usually troublesome, and simpler alternatives must be found. In practice, it is often easier to find the indefinite integral first and then the definite. The fundamental link is found in the following theorem:

If $g(x)$ is a function of x defined as the integral of a continuous function $f(x)$ in the form

$$g(x) \equiv \int f(x)\,dx,$$

then the differential coefficient $g'(x)$ exists, and

$$g'(x) = f(x).$$

In other words, $g(x)$ is a function whose differential coefficient is $f(x)$.

In order to prove this theorem, we go straight to the definition of a differential coefficient, and consider

$$g(x+h) - g(x).$$

If the lower limit of integration is a, then

$$\begin{aligned} g(x+h) - g(x) &= \int_a^{x+h} f(x)\,dx - \int_a^x f(x)\,dx \\ &= \int_x^{x+h} f(x)\,dx. \qquad \text{(Corollary (i), p. 83)} \end{aligned}$$

Now suppose that x_0 is a value of x at which $f(x)$ takes its least value in the interval $(x, x+h)$, and that x_1 is a value of x at which $f(x)$ takes its greatest value. Then, by the very definition of an integral, the value of $\int_x^{x+h} f(x)\,dx$ lies between $hf(x_0)$ and $hf(x_1)$—since h is the length of the interval of integration. Thus

$$\int_x^{x+h} f(x)\,dx$$

lies between $\quad hf(x_0), \quad hf(x_1),$

and so $\quad \dfrac{g(x+h) - g(x)}{h}$

lies between $\quad f(x_0), \quad f(x_1).$

Now take h to be progressively smaller and smaller. The values x_0, x_1 ultimately coincide with x itself, so that, since $f(x)$ is continuous,

$$\lim_{h \to 0} \frac{g(x+h) - g(x)}{h}$$

lies between two values which in the limit are each $f(x)$. Hence

$$g'(x) \equiv \lim_{h \to 0} \frac{g(x+h)-g(x)}{h}$$

exists, and its value is $f(x)$.

Thus $$g'(x) = f(x).$$

We therefore have the rule:

In order to integrate the function $f(x)$, we must find a function $g(x)$ of which it is the differential coefficient. (*Note the implication that $g(x)$ is continuous; compare the footnote on p.* 61.)

Note. If C is any constant,

$$\frac{d}{dx}\{g(x)+C\} = g'(x),$$

so that $g(x)$ is undetermined to the extent of an additive constant (compare p. 86).

Finally, we prove that *the integral*

$$g(x) \equiv \int f(x)\,dx$$

as evaluated by the rule $\quad g'(x) = f(x)$

is unique apart from the additive constant.

Suppose that, on the contrary, $h(x)$ is another function such that

$$h'(x) = f(x).$$

Write $$u(x) = g(x) - h(x),$$

where $u(x)$ is continuous since $g(x)$, $h(x)$ are.

Then
$$\begin{aligned} u'(x) &= g'(x) - h'(x) \\ &= f(x) - f(x) \\ &= 0. \end{aligned}$$

Hence (p. 62) the function $u(x)$ is constant.

Another way of stating this result is that *the equation*

$$\frac{dy}{dx} = f(x)$$

leads to the result $\quad y = \int f(x)\,dx + C$

uniquely.

7. The link with differentials. Suppose that y is a function of x whose differential coefficient is the given function $f(x)$, so that

$$\frac{dy}{dx} = f(x).$$

The differentials dy, dx satisfy (p. 43) the relation

$$dy = \frac{dy}{dx}dx,$$

or

$$dy = f(x)\,dx.$$

We have also just proved that

$$y = \int f(x)\,dx.$$

Hence we have the suggestive remark that the relation

$$dy = f(x)\,dx$$

between the differentials leads to the integrated relation

$$y = \int f(x)\,dx.$$

This is the justification for the common sequence of argument:

$$\frac{dy}{dx} = f(x);$$

therefore

$$dy = f(x)\,dx,$$

therefore

$$y = \int f(x)\,dx.$$

8. The evaluation of a definite integral. Once we have obtained (by methods to be given later) the indefinite integral

$$\int f(x)\,dx$$

by finding a function $g(x)$ whose differential coefficient is $f(x)$, we can evaluate the definite integral

$$\int_a^b f(x)\,dx$$

by a simple rule. For suppose that

$$F(x) = \int_a^x f(x)\,dx.$$

Then
$$\frac{d}{dx}\{F(x)-g(x)\}$$
$$= f(x)-f(x) \qquad \text{[p. 87 and definition of } g(x)]$$
$$= 0,$$

so that (p. 62) $\quad F(x)-g(x) = \text{constant}.$

But (p. 83) $F(x) = 0$ when $x = a$, and the value of the constant is therefore $-g(a)$, so that

$$F(x) = g(x)-g(a).$$

Hence
$$\int_a^b f(x)\,dx = g(b)-g(a),$$

each side being equal to $F(b)$.

Thus we have the rule:

In order to evaluate the definite integral

$$\int_a^b f(x)\,dx,$$

find a function $g(x)$ whose differential coefficient is $f(x)$; then

$$\int_a^b f(x)\,dx = g(b)-g(a).$$

NOTATION. It is often convenient to write

$$\Big[g(x)\Big]_a^b$$

to denote $\quad g(b)-g(a).$

9. Some simple standard forms. The evaluation of

$$\int f(x)\,dx$$

by finding a function $g(x)$ whose differential coefficient is $f(x)$ is naturally a matter of some difficulty. A start can, however, be made by inverting the formulæ for differentiation given on p. 37.

This gives us the formulæ:

$$\int x^n\,dx = \frac{x^{n+1}}{n+1} + C \quad (n \neq -1).$$

$$\int \cos x\,dx = \sin x + C.$$

$$\int \sin x\,dx = -\cos x + C$$

$$\int \sec x \tan x\,dx = \sec x + C.$$

$$\int \sec^2 x\,dx = \tan x + C.$$

$$\int \operatorname{cosec} x \cot x\,dx = -\operatorname{cosec} x + C.$$

$$\int \operatorname{cosec}^2 x\,dx = -\cot x + C.$$

We also have (pp. 40, 41)

$$\int \frac{dx}{\sqrt{(1-x^2)}} = \sin^{-1} x + C.$$

$$\int \frac{dx}{1+x^2} = \tan^{-1} x + C.$$

In the case of $\int \frac{dx}{\sqrt{(1-x^2)}}$, the arbitrary constant may be taken as a multiple of $\frac{1}{2}\pi$ and so the ambiguity of sign may be allowed for.

Note. The following results are also easily proved. If A is a constant, then

$$\int A f(x)\,dx = A \int f(x)\,dx;$$

and if $f(x), g(x)$ are integrable functions of x, then

$$\int \{f(x) + g(x)\}\,dx = \int f(x)\,dx + \int g(x)\,dx.$$

ILLUSTRATION 2. *To find the height at time t of a particle projected vertically upwards under gravity with speed V.*

The physical law is that a particle moving freely under gravity is subject to an acceleration vertically downwards of amount usually denoted by g. In ordinary foot-second units, the value of g is approximately 32.

Let x be the height of the particle above the ground at time t. Then (p. 48) its acceleration (vertically *upwards*) is $\frac{d^2x}{dt^2}$, so that

$$\frac{d^2x}{dt^2} = -g,$$

or

$$\frac{d(\dot{x})}{dt} = -g \quad \left(\dot{x} \equiv \frac{dx}{dt}\right).$$

Hence, integrating to find the function $\dot{x}$,

$$\dot{x} = -gt + C.$$

Now we are given that the velocity $\dot{x}$ is equal to V when t is zero, so that $C = V$. Hence

$$\frac{dx}{dt} = -gt + V.$$

Integrate to find x; then

$$x = -\tfrac{1}{2}gt^2 + Vt + A.$$

But $x = 0$ when $t = 0$ if the origin for x is taken at the point of projection, and so $A = 0$. Hence

$$x = Vt - \tfrac{1}{2}gt^2.$$

EXAMPLES

1. Prove by subdivision of the interval and actual summation that

$$\text{(i)} \int_1^2 dx = 1, \quad \text{(ii)} \int_0^3 x\,dx = \tfrac{9}{2}, \quad \text{(iii)} \int_2^3 x\,dx = \tfrac{5}{2}.$$

2. Find the indefinite integrals

$$\int x^3\,dx, \quad \int \frac{dx}{x^3}, \quad \int x^5\,dx.$$

3. Find the indefinite integrals

$$\int \cos 2x\,dx, \quad \int \sin \tfrac{1}{2}x\,dx, \quad \int 2\cos^2 \tfrac{1}{2}x\,dx.$$

4. Write out a formal proof of the theorems

(i) $\int_a^b Af(x)\,dx = A\int_a^b f(x)\,dx,$

(ii) $\int_a^b \{f(x)+g(x)\}\,dx = \int_a^b f(x)\,dx + \int_a^b g(x)\,dx$

based on the definition of § 2 (p. 81).

5. Evaluate the definite integrals

$$\int_1^3 x^2\,dx, \quad \int_{-1}^1 x^3\,dx, \quad \int_0^1 5x^4\,dx, \quad \int_2^4 (x^3+x)\,dx.$$

6. Evaluate the definite integrals

$$\int_0^{\frac{1}{2}\pi} \cos x\,dx, \quad \int_0^{\frac{1}{2}\pi} \sin x\,dx, \quad \int_0^{2\pi} \sin^2 \tfrac{1}{2}x\,dx, \quad \int_0^1 \frac{dx}{1+x^2}.$$

7. Evaluate the definite integrals

$$\int_0^1 (1-x)^2\,dx, \quad \int_1^2 (x-1)(2-x)\,dx, \quad \int_1^3 \left(x^2+\frac{1}{x^2}\right)^2 dx.$$

8. Find the value of

$$\int_0^{\frac{1}{2}\pi} \sin^2 x\,dx + \int_0^{\frac{1}{2}\pi} \cos^2 x\,dx.$$

CHAPTER V

DEVICES IN INTEGRATION

THE DIRECT EVALUATION of $\int f(x)\,dx$ by finding a function $g(x)$ with differential coefficient $f(x)$ is possible only in the simplest cases. A number of methods are, however, available to extend our scope considerably.

1. Substitution, or change of variable; indefinite integrals.

Consider an integral such as

$$I \equiv \int \frac{x^4\,dx}{(x^5+1)^3}.$$

Experience (gained by solving many similar examples) enables us to see that the 'essential variable' is not really x, but x^5+1. We therefore begin by finding the effect of replacing x^5+1 by a new variable t, so that

$$t = x^5+1.$$

Then

$$dt = 5x^4\,dx,$$

so that

$$I = \frac{1}{5}\int\frac{dt}{t^3} = -\frac{1}{10}t^{-2}$$

$$= \frac{-1}{10(x^5+1)^2}.$$

The *substitution* of t for the function x^5+1 has therefore led to the evaluation of the integral.

As another example, consider

$$I \equiv \int \sin^7 x \cos x\,dx.$$

The 'essential variable' is $\sin x$, so we write

$$t = \sin x,$$

giving

$$dt = \cos x\,dx.$$

Hence $$I = \int t^7\,dt = \tfrac{1}{8}t^8$$
$$= \tfrac{1}{8}\sin^8 x.$$

The theoretical basis on which the justification for the method of substitution rests is as follows:

Let $$I \equiv \int f(x)\,dx$$

be a given integral, and suppose that a certain function $u(x)$ appears likely as an 'essential variable' for the integration. Write

$$t = u(x).$$

Then $$\frac{dI}{dx} = f(x) \quad \text{(p. 87)}.$$

But $$\frac{dI}{dx} = \frac{dI}{dt}\frac{dt}{dx} \quad \text{(p. 27)},$$

so that $$f(x) = \frac{dI}{dt}u'(x),$$

or $$\frac{dI}{dt} = \frac{f(x)}{u'(x)}.$$

But the relation $t = u(x)$ enables us to express x in terms of t, and so $f(x)/\{u'(x)\}$ may be expressed as a function of t, say $F(t)$. Then

$$\frac{dI}{dt} = F(t),$$

so that (p. 89) $$I = \int F(t)\,dt.$$

Thus the effect of the substitution

$$t = u(x)$$

is (i) to replace $u'(x)dx$ by dt; (ii) to replace $f(x)/\{u'(x)\}$ by the corresponding function $F(t)$ of t; (iii) to replace integration with respect to x by integration with respect to t.

Note. The expression $\dfrac{f(x)}{u'(x)}$ may look complicated; but the whole virtue of the method lies in the selection of the function $u(x)$ in such a way that $f(x)/\{u'(x)\}$ falls naturally into the form

$F(t)$ as a function of t. There is no point in remembering this formula; it is the method of substitution that must be known.

COROLLARY. *To prove that, if a, b are constants, then*

$$\int f(ax+b)\,dx = \frac{1}{a}\int f(t)\,dt.$$

Write $t = ax+b$, so that $dt = a\,dx$. Then

$$\int f(ax+b)\,dx = \int f(t)\frac{dt}{a} = \frac{1}{a}\int f(t)\,dt.$$

For example,
$$\int \sin 5x\,dx = -\tfrac{1}{5}\cos 5x.$$

After a little practice, the reader should find that he uses this corollary so automatically that he is unaware of any substitution at all.

EXAMPLES I

Find the following integrals by performing the substitutions *mentally*:

1. $\int \sin 2x\,dx.$ 2. $\int \cos 3x\,dx.$ 3. $\int \sin \tfrac{1}{2}x\,dx.$

4. $\int \sec^2 4x\,dx.$ 5. $\int \sec \tfrac{1}{2}x \tan \tfrac{1}{2}x\,dx.$ 6. $\int (x+1)^2\,dx.$

7. $\int (x+3)^3\,dx.$ 8. $\int (x+5)^4\,dx.$ 9. $\int \frac{dx}{(x+1)^4}.$

10. $\int \frac{dx}{(x-2)^5}.$ 11. $\int \frac{dx}{(x-1)^3}.$ 12. $\int \frac{dx}{(x+5)^2}.$

13. $\int (2x+1)^2\,dx.$ 14. $\int (5x-3)^3\,dx.$ 15. $\int (\tfrac{1}{2}x+7)^4\,dx.$

16. $\int \frac{dx}{(2x+1)^2}.$ 17. $\int \frac{dx}{(4x-3)^3}.$ 18. $\int \frac{dx}{(\tfrac{1}{3}x-1)^4}.$

19. $\int \sqrt{(x+1)}\,dx.$ 20. $\int (x-1)^{\frac{3}{2}}\,dx.$ 21. $\int \frac{dx}{\sqrt{(x+1)}}.$

22. $\int \frac{dx}{\sqrt{(2x+1)}}.$ 23. $\int \frac{dx}{(2x-3)^{\frac{3}{2}}}.$ 24. $\int \frac{dx}{(3x+1)^{\frac{1}{3}}}.$

Find the following integrals:

25. $\int \frac{x\,dx}{(x^2+1)^2}$.
26. $\int \frac{x^2\,dx}{(x^3+1)^4}$.
27. $\int \frac{x^4\,dx}{(3x^5+1)^2}$.
28. $\int \frac{x\,dx}{(1-5x^2)^3}$.
29. $\int \cos^5 x \sin x\,dx$.
30. $\int \sin^3 x\,dx$.
31. $\int \sin^2 x \cos^3 x\,dx$.
32. $\int \tan^2 3x\,dx$.
33. $\int \frac{\sec^2 x\,dx}{(1+\tan x)^3}$.
34. $\int (1+\sec x)\sec x \tan x\,dx$.
35. $\int \sec^4 x\,dx$.
36. $\int \sec^2 x \tan^2 x\,dx$.

Note. It is sometimes more convenient to make the substitution from x to t in the alternative form

$$x = v(t).$$

We then have

$$\int f(x)\,dx = \int f\{v(t)\}\,v'(t)\,dt,$$

an integral with respect to t. This is, of course, easier to state than the earlier form (p. 95), but in practice the substitution $t = u(x)$ is probably the commoner.

ILLUSTRATION 1. *To find*

$$I \equiv \int \frac{dx}{(x^2+9)^2}.$$

Let $$x = 3\tan\theta$$

(remembering that $\tan^2\theta + 1 = \sec^2\theta$),

so that $$dx = 3\sec^2\theta\,d\theta.$$

Then
$$\begin{aligned} I &= \int \frac{3\sec^2\theta\,d\theta}{(9\tan^2\theta+9)^2} = \int \frac{3\sec^2\theta\,d\theta}{81\sec^4\theta} \\ &= \frac{1}{27}\int \cos^2\theta\,d\theta \\ &= \frac{1}{54}\int 2\cos^2\theta\,d\theta = \frac{1}{54}\int (1+\cos 2\theta)\,d\theta \\ &= \frac{1}{54}(\theta + \tfrac{1}{2}\sin 2\theta) = \frac{1}{54}(\theta + \sin\theta\cos\theta). \end{aligned}$$

Now $$\frac{\sin\theta}{x} = \frac{\cos\theta}{3} = \frac{1}{\sqrt{(x^2+9)}}.$$

Hence $$I = \frac{1}{54}\left\{\tan^{-1}(\tfrac{1}{3}x) + \frac{3x}{x^2+9}\right\}.$$

EXAMPLES II

Find the following integrals:

1. $\displaystyle\int\frac{dx}{(x^2+4)^2}$.

2. $\displaystyle\int\frac{x^2\,dx}{\sqrt{(1-x^2)}}$.

3. $\displaystyle\int\frac{x^2\,dx}{1+x^2}$.

4. $\displaystyle\int\sqrt{(1-x^2)}\,dx$.

5. $\displaystyle\int\frac{1+x}{\sqrt{(1-x^2)}}\,dx$.

6. $\displaystyle\int x^2\sqrt{(1-x^2)}\,dx$.

[For examples 7–10, compare the device at the start of Illustration 4, p. 100.]

7. $\displaystyle\int\frac{x\,dx}{\sqrt{(1-x)}}$.

8. $\displaystyle\int\frac{x\,dx}{\sqrt{(9-x)}}$.

9. $\displaystyle\int x\sqrt{(1-x)}\,dx$.

10. $\displaystyle\int x^2\sqrt{(x-4)}\,dx$.

2. Substitution, or change of variable; definite integrals. The introduction of limits for the integral presents no essentially new feature, except, of course, that the original limits for x must now be replaced by corresponding limits for the new variable t. (However, we add the warning that care must be taken in more advanced examples.)

ILLUSTRATION 2. *To evaluate*

$$\int_0^{\frac{1}{2}\pi}\sin^4 x\cos x\,dx.$$

Write $$t = \sin x;$$

then $$dt = \cos x\,dx.$$

Hence $$\int\sin^4 x\cos x\,dx = \int t^4\,dt.$$

Now as x, starting from zero, increases to $\frac{1}{2}\pi$, the new variable t, starting from zero, increases to 1. Hence

$$\int_0^{\frac{1}{2}\pi} \sin^4 x \cos x\, dx = \int_0^1 t^4\, dt$$

$$= \left[\frac{1}{5}t^5\right]_0^1$$

$$= \frac{1}{5}.$$

ILLUSTRATION 3. *To evaluate.*

$$\int_2^8 \frac{dx}{(5x+2)^3}.$$

Let $$t = 5x+2;$$

then $$dt = 5dx.$$

Hence $$\int \frac{dx}{(5x+2)^3} = \frac{1}{5}\int \frac{dt}{t^3}.$$

Now as x, starting from 2, increases to 8, the new variable t, starting from 12, increases to 42. Hence

$$\int_2^8 \frac{dx}{(5x+2)^3} = \frac{1}{5}\int_{12}^{42} \frac{dt}{t^3}$$

$$= -\frac{1}{10}\left[\frac{1}{t^2}\right]_{12}^{42}$$

$$= -\frac{1}{10}\left[\frac{1}{42^2} - \frac{1}{12^2}\right]$$

$$= -\frac{1}{360}\left[\frac{1}{7^2} - \frac{1}{2^2}\right]$$

$$= -\frac{1}{360}\left[\frac{-45}{196}\right]$$

$$= \frac{1}{1568}.$$

ILLUSTRATION 4. *To evaluate*

$$\int_3^{11} \frac{dx}{(x+5)\sqrt{(1+x)}}.$$

It is a common device in such integrals to write t^2 for the expression under the square root. Thus

$$x = t^2 - 1,$$

so that
$$dx = 2t\,dt.$$

Also, as x rises from 3 to 11, t rises from 2 to $2\sqrt{3}$. Hence

$$\int_3^{11} \frac{dx}{(x+5)\sqrt{(1+x)}} = \int_2^{2\sqrt{3}} \frac{2t\,dt}{(t^2+4)t} = 2\int_2^{2\sqrt{3}} \frac{dt}{t^2+4}.$$

Now write
$$t = 2\tan\theta,$$

so that
$$dt = 2\sec^2\theta\,d\theta.$$

As t rises from 2 to $2\sqrt{3}$, θ rises from $\frac{1}{4}\pi$ to $\frac{1}{3}\pi$. Hence

$$I = 2\int_{\frac{1}{4}\pi}^{\frac{1}{3}\pi} \frac{2\sec^2\theta\,d\theta}{4\tan^2\theta+4} = \int_{\frac{1}{4}\pi}^{\frac{1}{3}\pi} d\theta$$

$$= \Big[\theta\Big]_{\frac{1}{4}\pi}^{\frac{1}{3}\pi} = \frac{1}{3}\pi - \frac{1}{4}\pi$$

$$= \frac{1}{12}\pi.$$

ILLUSTRATION 5. *To find the area of a circle of radius a.*

Referred to rectangular Cartesian axes with the origin at the centre, the equation of the circle (Fig. 34) is

$$x^2 + y^2 = a^2,$$

so that
$$y = \pm\sqrt{(a^2 - x^2)},$$

the two values of y for a given value of x corresponding to the parts of the circle 'above' and 'below' the x-axis.

The area of the whole circle is double that of the upper semi-circle, and so we may take

$$A = 2\int_{-a}^{a} \sqrt{(a^2 - x^2)}\,dx.$$

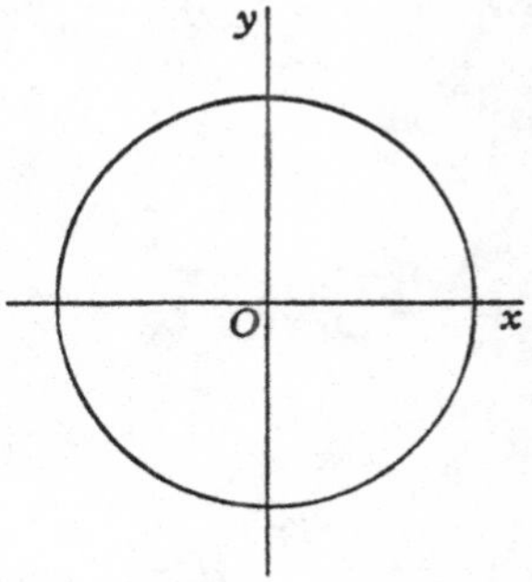

Fig. 34.

Make the substitution $x = a \sin t,$

so that $dx = a \cos t\, dt;$

the range $(-a, a)$ of x corresponds to the range $(-\frac{1}{2}\pi, \frac{1}{2}\pi)$ of t. Hence

$$A = 2\int_{-\frac{1}{2}\pi}^{\frac{1}{2}\pi} a \cos t \,.\, a \cos t\, dt.$$

[*Note.* Since $a \cos t$ is *positive* in the range $(-\frac{1}{2}\pi, \frac{1}{2}\pi)$, we have taken the positive square root throughout the range.]

Thus
$$A = a^2 \int_{-\frac{1}{2}\pi}^{\frac{1}{2}\pi} 2\cos^2 t\, dt = a^2 \int_{-\frac{1}{2}\pi}^{\frac{1}{2}\pi} (1 + \cos 2t)\, dt$$

$$= a^2 \Big[t + \tfrac{1}{2}\sin 2t \Big]_{-\frac{1}{2}\pi}^{\frac{1}{2}\pi}$$

$$= a^2\{(\tfrac{1}{2}\pi) - (-\tfrac{1}{2}\pi)\}$$

$$= \pi a^2.$$

EXAMPLES III

Evaluate the following definite integrals:

1. $\displaystyle\int_0^{\frac{1}{2}\pi} \sin^2\theta \cos\theta\, d\theta.$

2. $\displaystyle\int_0^{\frac{1}{2}\pi} \sin^3\theta\, d\theta.$

3. $\displaystyle\int_0^1 \frac{dx}{(1+x^2)^2}.$

4. $\displaystyle\int_0^1 \sqrt{(1-x^2)}\, dx.$

5. $\displaystyle\int_0^{\frac{1}{2}} \sqrt{(1-x^2)}\, dx.$

6. $\displaystyle\int_0^{\frac{1}{4}\pi} \sec^2\theta \tan^2\theta\, d\theta.$

7. $\displaystyle\int_0^{\frac{1}{4}\pi} \sec^4\theta\, d\theta.$

8. $\displaystyle\int_{\frac{1}{3}\pi}^{\frac{1}{2}\pi} \cos^5\theta\, d\theta.$

9. $\displaystyle\int_0^1 \sqrt{(1-x)}\, dx.$

10. $\displaystyle\int_1^4 \sqrt{(5-x)}\, dx.$

ILLUSTRATION 6.* *To evaluate*

$$\int_0^{\pi} \frac{\sin\theta\, d\theta}{\sqrt{(1 - 2a\cos\theta + a^2)}}.$$

It is implicit in such a question that the *positive* square root is to be taken.

* Illustration 6 may be postponed, if desired.

Let $$t = +\sqrt{(1-2a\cos\theta+a^2)},$$

so that $$t^2 = 1-2a\cos\theta+a^2,$$

and $$2t\,dt = 2a\sin\theta\,d\theta.$$

Hence $$\int\frac{\sin\theta\,d\theta}{\sqrt{(1-2a\cos\theta+a^2)}} = \frac{1}{a}\int\frac{t\,dt}{t} = \frac{1}{a}\int dt$$
$$= t/a.$$

Consider now the range of values of t as θ runs from 0 to π. When $\theta = 0$, we have the relation

$$\sqrt{(1-2a\cos\theta+a^2)} = \sqrt{(1-2a+a^2)} = \pm(1-a),$$

that sign being taken which makes the square root positive. When $\theta = \pi$, we have

$$\sqrt{(1-2a\cos\theta+a^2)} = \sqrt{(1+2a+a^2)} = \pm(1+a),$$

that sign again being taken which makes the square root positive. Various cases must therefore be considered:

(i) *When a lies between* $-1, 1$. The positive square roots are $1-a$, $1+a$ respectively, so that the value of the integral is

$$\frac{1}{a}\Big[t\Big]_{1-a}^{1+a}$$
$$= \frac{1}{a}\{(1+a)-(1-a)\} = 2.$$

The value of t varies continuously from $1-a$, through 1 (when $\cos\theta = \frac{1}{2}a$), to $1+a$ as θ moves from 0 to π.

(ii) *When a is greater than* 1. The positive square roots are $a-1$, $a+1$ respectively, so that the value of the integral is

$$\frac{1}{a}\Big[t\Big]_{a-1}^{a+1}$$
$$= \frac{1}{a}\{(a+1)-(a-1)\} = \frac{2}{a}.$$

The value of t rises continuously from $a-1$ to $a+1$ as θ moves from 0 to π.

(iii) *When a (being negative) is less than* -1. The positive square roots are $1-a$ and $-(1+a)$ respectively. (For example, if $a=-4$, the square roots are $1-(-4)$ and $-(1-4)$, that is to say, 5 and 3.) Hence the value of the integral is

$$\frac{1}{a}\Big[t\Big]_{1-a}^{-1-a}$$
$$=\frac{1}{a}\{(-1-a)-(1-a)\}=-\frac{2}{a}.$$

This value looks negative, but is actually positive, since a is negative.

The value of t falls continuously from $1-a$ to $-1-a$ as θ moves from 0 to π.

SUMMARY. The value of the integral is $2, 2/a, -2/a$ according as $-1<a<1, a>1, a<-1$.

EXAMPLES IV

Evaluate the integrals:

1. $\displaystyle\int_0^\pi \frac{\sin\theta\,d\theta}{\surd(5-4\cos\theta)}$. 2. $\displaystyle\int_0^\pi \frac{\sin\theta\,d\theta}{\surd(10+6\cos\theta)}$. 3. $\displaystyle\int_0^\pi \frac{\sin\theta\,d\theta}{\surd(50-14\cos\theta)}$.

3. Integration by parts.

A very important method of integration follows as a direct result of integrating the formula for the differentiation of a product. If $u(x), v(x)$ are given functions of x, then

$$\frac{d}{dx}(uv)=u\frac{dv}{dx}+v\frac{du}{dx},$$

or

$$v\frac{du}{dx}=\frac{d}{dx}(uv)-u\frac{dv}{dx}.$$

Hence, on integrating from an arbitrary lower limit a, and denoting the variable of integration by the letter t (p. 85), we obtain the important formula

$$\int_a^x v(t)\,u'(t)\,dt=\Big[u(t)\,v(t)\Big]_a^x-\int_a^x u(t)\,v'(t)\,dt.$$

We observe that the left-hand side is the integral of the product of two functions v, u', *one of which can be recognized as the differential coefficient of a function u.*

It is customary to quote this result in its 'indefinite integral' form

$$\int v(x)\,u'(x)\,dx=u(x)\,v(x)-\int u(x)\,v'(x)\,dx,$$

and we shall usually do so (Illustrations 7, 8); but note the warning which follows.

Warning. The omission of the term $-u(a)\,v(a)$ may occasionally lead to curious results. For example, if $v(x) \equiv \sec x$, $u(x) \equiv \cos x$, we appear to have

$$\int \sec x(-\sin x)\,dx = \cos x \sec x - \int \cos x(\sec x \tan x)\,dx,$$

or

$$-\int \tan x\,dx = 1 - \int \tan x\,dx,$$

or

$$0 = 1.$$

(The product $\cos x \sec x$ is a constant rather than a genuine function of x, and would be cancelled in *definite* integration.)

ILLUSTRATION 7. *To find*

$$\int x \sin x\,dx.$$

We have to integrate the product $x \sin x$, where we recognize $\sin x$ as the differential coefficient of the function $-\cos x$. Applying the formula, we have

$$\begin{aligned}\int x \sin x\,dx &= \int x \frac{d}{dx}(-\cos x)\,dx \\ &= x(-\cos x) - \int (-\cos x).1.dx \\ &= -x \cos x + \int \cos x\,dx \\ &= -x \cos x + \sin x,\end{aligned}$$

the desired value.

It may be remarked that an alternative first step could be

$$\int \sin x \frac{d}{dx}(\tfrac{1}{2}x^2)\,dx$$

leading to

$$\begin{aligned}&\sin x.\tfrac{1}{2}x^2 - \int \tfrac{1}{2}x^2 \frac{d}{dx}(\sin x)\,dx \\ &= \tfrac{1}{2}x^2 \sin x - \tfrac{1}{2}\int x^2 \cos x\,dx.\end{aligned}$$

However, a glance at the integral

$$\int x^2 \cos x\,dx$$

will convince the reader that he has not made the problem any simpler by this beginning.

The whole success of the method depends on an intelligent first step, and certainty of touch should be acquired by working many examples.

ILLUSTRATION 8. *To find*

$$I \equiv \int \sin^{-1} x\, dx.$$

This example illustrates a case where the possibility of integration by parts might pass unsuspected. We have

$$I = \int \sin^{-1} x \,.\, 1 \,.\, dx = \int \sin^{-1} x \,.\, \frac{d(x)}{dx}\, dx$$

$$= x \sin^{-1} x - \int x \,.\, \frac{1}{\sqrt{(1-x^2)}}\, dx.$$

Write $$J = \int \frac{x\, dx}{\sqrt{(1-x^2)}},$$

and put $$1 - x^2 = y^2,$$

so that $$-2x\, dx = 2y\, dy.$$

Then $$J = \int \frac{-y\, dy}{y} = -\int dy = -y$$

$$= -\sqrt{(1-x^2)}.$$

Hence $$I = x \sin^{-1} x + \sqrt{(1-x^2)}.$$

EXAMPLES V

Find the integrals:

1. $\int x \cos x\, dx.$
2. $\int x^2 \cos x\, dx.$
3. $\int 2x \sec^2 x \tan x\, dx.$
4. $\int x \tan^{-1} x\, dx.$

Evaluate:

5. $\int_0^{\frac{1}{2}\pi} x \cos x\, dx.$
6. $\int_0^{\frac{1}{2}\pi} x^2 \sin x\, dx.$
7. $\int_0^{\pi} (1+x)^2 \sin x\, dx.$
8. $\int_0^{\frac{1}{2}\pi} (x-2)^2 \cos x\, dx.$

In conclusion, we remind the reader of the value of a good technique in the normal processes of algebra and trigonometry For example,

$$\int \sin 3x \sin 5x \, dx = \frac{1}{2}\int \{\cos 2x - \cos 8x\} \, dx$$
$$= \frac{1}{4}\sin 2x - \frac{1}{16}\sin 8x.$$

Again,
$$\int \sin^4 x \, dx = \frac{1}{4}\int (2\sin^2 x)^2 \, dx$$
$$= \frac{1}{4}\int (1-\cos 2x)^2 \, dx$$
$$= \frac{1}{4}\int (1 - 2\cos 2x + \cos^2 2x) \, dx$$
$$= \frac{1}{4}\int \left\{1 - 2\cos 2x + \frac{1}{2}(1+\cos 4x)\right\} dx$$
$$= \frac{1}{4}\int \left(\frac{3}{2} - 2\cos 2x + \frac{1}{2}\cos 4x\right) dx$$
$$= \frac{1}{4}\left\{\frac{3}{2}x - \sin 2x + \frac{1}{8}\sin 4x\right\}.$$

EXAMPLES VI

Find the integrals:

1. $\int \sin 5x \cos x \, dx.$

2. $\int \cos^2 x \, dx.$

3. $\int \sin^2 4x \, dx.$

4. $\int \sin 4x \sin 2x \, dx.$

Evaluate:

5. $\int_0^{\frac{1}{2}\pi} \sin x \cos 4x \, dx.$

6. $\int_0^{\pi} \sin^3 x \, dx.$

7. $\int_0^{\pi} \cos x \cos \tfrac{1}{2}x \, dx.$

8. $\int_0^{\frac{1}{2}\pi} \cos^4 x \, dx.$

4.* Formulæ of reduction.

Consider the integral

$$I \equiv \int \sin^n x \, dx \quad (n \text{ a positive integer}).$$

* This paragraph may be postponed, if desired.

Writing this in the form

$$I \equiv \int \sin^{n-1} x \, . \sin x \, dx,$$

and integrating by parts, we have

$$I = -\sin^{n-1} x \cos x + \int (n-1) \sin^{n-2} x \cos x \, . \cos x \, dx$$

$$= -\sin^{n-1} x \cos x + (n-1) \int \sin^{n-2} x (1 - \sin^2 x) \, dx$$

$$= -\sin^{n-1} x \cos x + (n-1) \int \sin^{n-2} x \, dx - (n-1) \int \sin^n x \, dx$$

$$= -\sin^{n-1} x \cos x + (n-1) \int \sin^{n-2} x \, dx - (n-1) I.$$

Hence $$nI = -\sin^{n-1} x \cos x + (n-1) \int \sin^{n-2} x \, dx,$$

or $$I = -\frac{\sin^{n-1} x \cos x}{n} + \frac{n-1}{n} \int \sin^{n-2} x \, dx.$$

In this way we have made the integral of $\sin^n x$ depend on the integral of $\sin^{n-2} x$, whose degree in $\sin x$ is less than that of the original. To emphasize this reduction of degree, we write

$$I_n \equiv \int \sin^n x \, dx, \quad I_{n-2} \equiv \int \sin^{n-2} x \, dx.$$

Thus $$I_n = -\frac{\sin^{n-1} x \cos x}{n} + \frac{n-1}{n} I_{n-2}.$$

It is implicit that $n \geqslant 2$.

In the same way,

$$I_{n-2} = -\frac{\sin^{n-3} x \cos x}{n-2} + \frac{n-3}{n-2} I_{n-4},$$

$$I_{n-4} = -\frac{\sin^{n-5} x \cos x}{n-4} + \frac{n-5}{n-4} I_{n-6},$$

and so on.

By repeated application of the formula, we reach a stage where we must evaluate

$$I_1 \equiv \int \sin x \, dx = -\cos x$$

when n is odd; or $$I_0 \equiv \int dx = x,$$

when n is even.

A formula such as

$$I_n = -\frac{\sin^{n-1}x\cos x}{n} + \frac{n-1}{n} I_{n-2}$$

which enables us to reduce the index n by successive stages is called a FORMULA OF REDUCTION for the integral.

ILLUSTRATION 9. *To evaluate*

$$\int_0^{\frac{1}{2}\pi} \cos^{2n} x\, dx.$$

Write

$$I_n \equiv \int_0^{\frac{1}{2}\pi} \cos^{2n} x\, dx.$$

(This is a slight change of notation from the preceding, but perhaps a little more convenient.)

We have

$$\begin{aligned} I_n &\equiv \int_0^{\frac{1}{2}\pi} \cos^{2n-1} x \cos x\, dx \\ &= [\cos^{2n-1} x \sin x]_0^{\frac{1}{2}\pi} - \int_0^{\frac{1}{2}\pi} (2n-1)\cos^{2n-2} x(-\sin x).\sin x\, dx \\ &= 0 + (2n-1)\int_0^{\frac{1}{2}\pi} \cos^{2n-2} x(1-\cos^2 x)\, dx, \end{aligned}$$

since $\cos^{2n-1} x \sin x$ vanishes for $x = 0$, and, when $n \geqslant 1$, for $x = \frac{1}{2}\pi$.

Hence

$$I_n = (2n-1)(I_{n-1} - I_n),$$

or

$$I_n = \frac{2n-1}{2n} I_{n-1}.$$

Applying this formula successively, we have

$$\begin{aligned} I_n &= \frac{2n-1}{2n}\frac{2n-3}{2n-2} I_{n-2} \\ &= \frac{2n-1}{2n}\frac{2n-3}{2n-2}\frac{2n-5}{2n-4} I_{n-3} \\ &\cdots\cdots\cdots\cdots\cdots\cdots\cdots\cdots \\ &= \frac{2n-1}{2n}\frac{2n-3}{2n-2}\cdots\frac{3}{4} I_1 \\ &= \frac{2n-1}{2n}\frac{2n-3}{2n-2}\cdots\frac{3}{4}\frac{1}{2} I_0, \end{aligned}$$

where $$I_0 = \int_0^{\frac{1}{2}\pi} dx = \frac{\pi}{2}.$$

Hence $$I_n = \frac{(2n-1)(2n-3)\ldots 3.1}{2n(2n-2)\ldots 4.2}\frac{\pi}{2}.$$

EXAMPLES VII

Find the following integrals by the use of formulæ of reduction:

1. $\int \sin^6 x\,dx.$ 2. $\int \cos^7 x\,dx.$ 3. $\int \sin^9 x\,dx.$

Evaluate the following integrals by the use of formulæ of reduction:

4. $\int_0^{\frac{1}{2}\pi} \cos^8 x\,dx.$ 5. $\int_0^{\frac{1}{2}\pi} \sin^7 x\,dx.$ 6. $\int_0^{\frac{1}{2}\pi} \sin^{10} x\,dx.$

7. $\int_0^{\frac{1}{2}\pi} \sin^{2n} x\,dx.$ 8. $\int_0^{\frac{1}{2}\pi} \cos^{2n+1} x\,dx.$ 9. $\int_0^{\frac{1}{2}\pi} \sin^{2n+1} x\,dx.$

ILLUSTRATION 10. *To evaluate*

$$\int_0^{\frac{1}{2}\pi} \sin^{2m} x \cos^{2n+1} x\,dx.$$

To imply dependence on both m and n, write

$$I_{m,n} \equiv \int_0^{\frac{1}{2}\pi} \sin^{2m} x \cos^{2n+1} x\,dx.$$

Then $$I_{m,n} = \int_0^{\frac{1}{2}\pi} \sin^{2m} x \cos^{2n} x \cos x\,dx$$

$$= \int_0^{\frac{1}{2}\pi} \sin^{2m} x \cos^{2n} x\,d(\sin x)$$

$$= \left[\frac{1}{2m+1}\sin^{2m+1} x \cos^{2n} x\right]_0^{\frac{1}{2}\pi}$$

$$-\int_0^{\frac{1}{2}\pi} \frac{1}{2m+1}\sin^{2m+1} x\,.\,2n\cos^{2n-1} x(-\sin x)\,dx$$

$$= 0 + \frac{2n}{2m+1}\int_0^{\frac{1}{2}\pi} \sin^{2m+2} x \cos^{2n-1} x\,dx,$$

since $\sin^{2m+1}x\cos^{2n}x$ vanishes when $x = 0$, and, when $n > 0$, for $x = \frac{1}{2}\pi$. Hence

$$I_{m,n} = \frac{2n}{2m+1}\int_0^{\frac{1}{2}\pi} \sin^{2m}x\cos^{2n-1}x(1-\cos^2 x)\,dx$$

$$= \frac{2n}{2m+1}\{I_{m,n-1} - I_{m,n}\},$$

or
$$(2m+1+2n)I_{m,n} = 2nI_{m,n-1},$$

or
$$I_{m,n} = \frac{2n}{2m+2n+1}I_{m,n-1}.$$

Reducing n successively, we have

$$I_{m,n} = \frac{2n}{2m+2n+1}\frac{2n-2}{2m+2n-1}I_{m,n-2}$$

$$= \frac{2n}{2m+2n+1}\frac{2n-2}{2m+2n-1}\cdots\frac{2}{2m+3}I_{m,0}.$$

Also
$$I_{m,0} = \int_0^{\frac{1}{2}\pi}\sin^{2m}x\cos x\,dx$$

$$= \frac{1}{2m+1}\left[\sin^{2m+1}x\right]_0^{\frac{1}{2}\pi} = \frac{1}{2m+1}.$$

Hence
$$I_{m,n} = \frac{2n(2n-2)\dots 2}{(2m+2n+1)(2m+2n-1)\dots(2m+1)}.$$

EXAMPLES VIII

Evaluate the following integrals by the use of formulæ of reduction:

1. $\int_0^{\frac{1}{2}\pi}\sin^4 x\cos^7 x\,dx.$

2. $\int_0^{\frac{1}{2}\pi}\sin^6 x\cos^3 x\,dx.$

3. $\int_0^1 x^m(1-x)^n\,dx$ (by the substitution, $x = \sin^2\theta$).

4. $\int_0^{\frac{1}{2}\pi}\sin^{2m}x\cos^{2m}x\,dx.$

5. $\int_0^{\frac{1}{2}\pi}\sin^{2m+1}x\cos^{2m+1}x\,dx.$

6. $\int_0^1 x^6(1-x)^{\frac{5}{2}}\,dx.$

7. $\int_0^1 x^{\frac{3}{2}}(1-x)^{\frac{5}{2}}\,dx.$

CHAPTER VI

APPLICATIONS OF INTEGRATION

1. Use in dynamics. Suppose that a particle P (Fig. 35) is moving along a straight line so that at time t its distance from a

Fig. 35.

fixed point O of the line is x. We have seen that, if v, f are the velocity and acceleration respectively, then (p. 48)

$$v = \frac{dx}{dt},$$

$$f = \frac{dv}{dt} = \frac{d^2x}{dt^2}.$$

(i) If v is a known function of t, then x can be evaluated by means of the integral

$$x = \int v\,dt;$$

and if f is a known function of t, then v can be evaluated by means of the integral

$$v = \int f\,dt,$$

after which x can be determined by a further integration, as before.

At each integration an arbitrary constant is introduced; its value depends on given initial conditions for a particular problem.

(ii) If v is a known function of x, then t can be evaluated by means of the integral

$$t = \int \frac{dx}{v}.$$

If f is a known function of x, we find v by means of the relation (p. 27)

$$f = \frac{dv}{dt} = \frac{dv}{dx}.\frac{dx}{dt} = \frac{dv}{dx}.v,$$

so that
$$f = \frac{1}{2}\frac{d(v^2)}{dx}.$$

Hence
$$v^2 = \int 2f\,dx,$$

so that v can be calculated.

ILLUSTRATION 1. Suppose that f is given as a function of t by means of the formula
$$f = kt\cos t,$$

and that $v = k$, $x = 0$ when $t = 0$.

Then
$$\frac{dv}{dt} = kt\cos t,$$

so that
$$\begin{aligned} v &= k\int t\cos t\,dt \\ &= kt\sin t - k\int \sin t\,dt \quad \text{(by parts)} \\ &= kt\sin t + k\cos t + C, \end{aligned}$$

where, since $v = k$ when $t = 0$,
$$k = k + C,$$

or
$$C = 0.$$

Hence
$$v = kt\sin t + k\cos t;$$

that is,
$$\frac{dx}{dt} = kt\sin t + k\cos t,$$

so that
$$\begin{aligned} x &= k\int t\sin t\,dt + k\int \cos t\,dt \\ &= k\{-t\cos t + \sin t\} + k\sin t + D, \end{aligned}$$

where $D = 0$ since $x = 0$ when $t = 0$. Hence
$$x = k(2\sin t - t\cos t).$$

ILLUSTRATION 2. Suppose that f is given as a function of x by means of the formula
$$f = 2k^2x(x^2+1)$$

and that $x = 0$, $v = k$ when $t = 0$.

We have $$\frac{d(v^2)}{dx} = 4k^2x(x^2+1),$$

so that $$v^2 = k^2\int(4x^3+4x)\,dx$$

$$= k^2(x^4+2x^2)+C,$$

where, from the initial conditions,

$$k^2 = C.$$

Hence $$v^2 = k^2(x^2+1)^2,$$

so that $$v = \pm k(x^2+1).$$

But $v = k$ when $x = 0$, so the positive square root must be taken. Hence

$$v = k(x^2+1).$$

It follows that $$\frac{dx}{dt} = k(x^2+1),$$

so that $$kt = \int\frac{dx}{x^2+1}$$

$$= \tan^{-1}x + D.$$

But $x = 0$ when $t = 0$, so that $D = 0$, and

$$kt = \tan^{-1}x,$$

or $$x = \tan kt.$$

Note. The formulæ are relevant only up to time $\pi/2k$, when x tends to infinity.

2. Area; Cartesian coordinates. We have already (pp. 77–85) discussed in some detail the meaning and evaluation of the area 'under' the curve

$$y = f(x).$$

For completeness, we repeat the formula

$$\int_a^b f(x)\,dx$$

for the area $ABQP$ of the diagram (Fig. 36), reminding the reader to be careful about sign when the curve crosses the x-axis (p. 84).

It may be useful to note also that, for the area bounded by a curve

$$x = f(y),$$

two ordinates $y = c$, $y = d$, and the part of the y-axis between those ordinates (Fig. 37), the area is

$$\int_c^d f(y)\,dy.$$

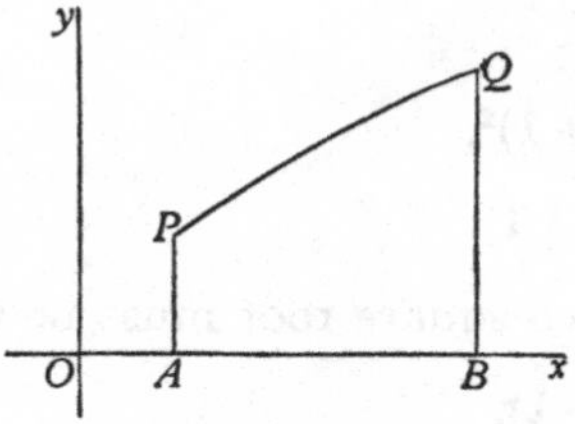

Fig. 36.

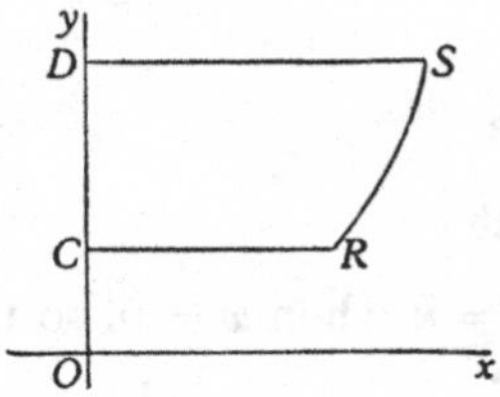

Fig. 37.

3. The area of a sector, in polar coordinates. *The equation of a curve in polar coordinates is*

$$r = f(\theta),$$

where r is a single-valued function of θ. To find an expression for the area of the sector OAB (Fig. 38), *bounded by the radii OA, OB given by $\theta = \alpha, \beta$ and the arc AB.*

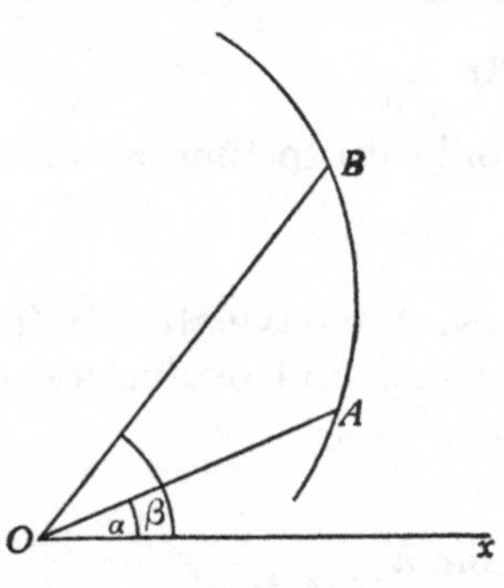

Fig. 38.

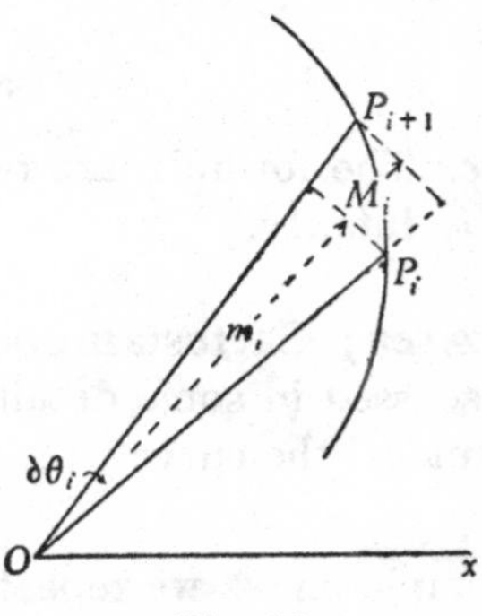

Fig. 39.

METHOD I. Divide the arc AB into n parts at points

$$A \equiv P_0, P_1, P_2, \ldots, P_{n-2}, P_{n-1}, P_n \equiv B.$$

Write $$\angle P_iOP_{i+1} = \delta\theta_i \text{ (Fig. 39)}.$$

Denote the least value of the function $f(\theta)$ in $\delta\theta_i$ by the symbol m_i and the greatest by M_i. Then the area of a typical element P_iOP_{i+1} lies between that of a *circular* sector of angle $\delta\theta_i$ and radius m_i, and that of a circular sector of angle $\delta\theta_i$ and radius M_i.

Moreover, we proved (p. 100) that the area of a circle of radius a is πa^2, so that, by proportion, the area of a sector of angle $\delta\theta_i$ is

$$\frac{\delta\theta_i}{2\pi}(\pi a^2)$$

or

$$\tfrac{1}{2}a^2\,\delta\theta_i.$$

Hence the area of the sector OAB lies between the two sums

$$\sum_0^{n-1}\tfrac{1}{2}m_i^2\,\delta\theta_i,\quad \sum_0^{n-1}\tfrac{1}{2}M_i^2\,\delta\theta_i.$$

In the limit, as the number of intervals increases indefinitely, the size of each interval decreasing indefinitely, these two sums, for 'ordinary' functions $f(\theta)$, approach (p. 82) the limit

$$\frac{1}{2}\int_\alpha^\beta\{f(\theta)\}^2 d\theta,$$

so that we have the formula

$$\text{Area } OAB = \frac{1}{2}\int_\alpha^\beta\{f(\theta)\}^2 d\theta$$

$$= \frac{1}{2}\int_\alpha^\beta r^2\,d\theta.$$

METHOD II. The area of a sector AOP (Fig. 40), where $\angle xOP = \theta$, is a function of the angle θ, say $A(\theta)$. Suppose that P′ is a point on the curve near to P, so that

$$\angle xOP' = \theta + \delta\theta;$$

then the area of the sector AOP' is $A(\theta+\delta\theta)$.

We make the postulate, in informal language, that, when P' is near to P, the area of the sector

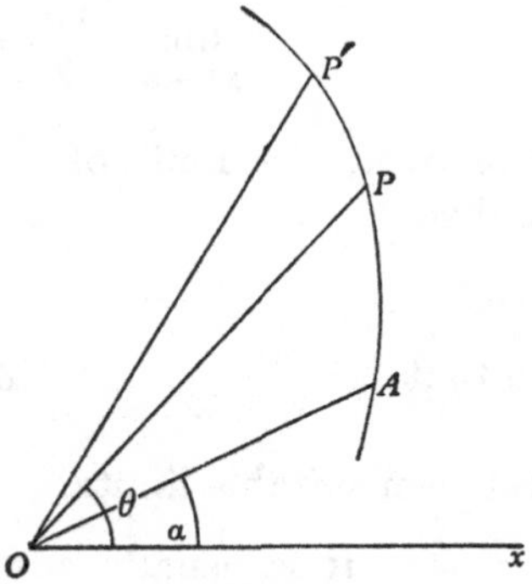

Fig. 40.

POP' is not very different from that of the triangle POP'; in more formal language, that

$$\lim_{P'\to P}\frac{\text{area of sector } POP'}{\text{area of triangle } POP'}=1.$$

Now $$\frac{A(\theta+\delta\theta)-A(\theta)}{\delta\theta}=\frac{\text{sector } POP'}{\delta\theta}$$

$$=\frac{\text{sector } POP'}{\Delta\, POP'}\cdot\frac{\Delta\, POP'}{\delta\theta},$$

where $\Delta\, POP'$ is the area of the triangle POP' so that

$$\Delta\, POP'=\tfrac{1}{2}OP\,.\,OP'\sin\delta\theta.$$

Hence

$$\frac{A(\theta+\delta\theta)-A(\theta)}{\delta\theta}=\frac{\text{sector } POP'}{\Delta\, POP'}\,.\,(\tfrac{1}{2}OP\,.\,OP').\frac{\sin\delta\theta}{\delta\theta}.$$

Let $\delta\theta\to 0$, so that $P'\to P$. Then

$$\lim_{\delta\theta\to 0}\frac{A(\theta+\delta\theta)-A(\theta)}{\delta\theta}=A'(\theta);$$

$$\lim_{P'\to P}\frac{\text{sector } POP'}{\Delta\, POP'}=1 \quad \text{(postulate)};$$

$$\lim_{P'\to P}(\tfrac{1}{2}OP\,.\,OP')=\tfrac{1}{2}OP^2=\tfrac{1}{2}r^2;$$

$$\lim_{\delta\theta\to 0}\frac{\sin\delta\theta}{\delta\theta}=1 \quad \text{(p. 32).}$$

Moreover, the limit of a product is the product of its limits, and so

$$A'(\theta)=\tfrac{1}{2}r^2,$$

so that $$A(\theta)=\int\tfrac{1}{2}r^2d\theta$$

between suitable limits.

Note. It is usually advisable to make a rough sketch of the curve before applying this formula.

ILLUSTRATION 3. *To find the area of the curve (a limaçon) for which* $r = 1 + \frac{1}{2}\cos\theta$.

The curve (Fig. 41) may be sketched by putting

$$\theta = 0, \tfrac{1}{6}\pi, \tfrac{1}{3}\pi, \tfrac{1}{2}\pi, \ldots, 2\pi$$

in succession; note that

$$\frac{dr}{d\theta} = -\tfrac{1}{2}\sin\theta,$$

so that r decreases steadily between $0, \pi$, after which it increases again. The curve is symmetrical about the line $\theta = 0$.

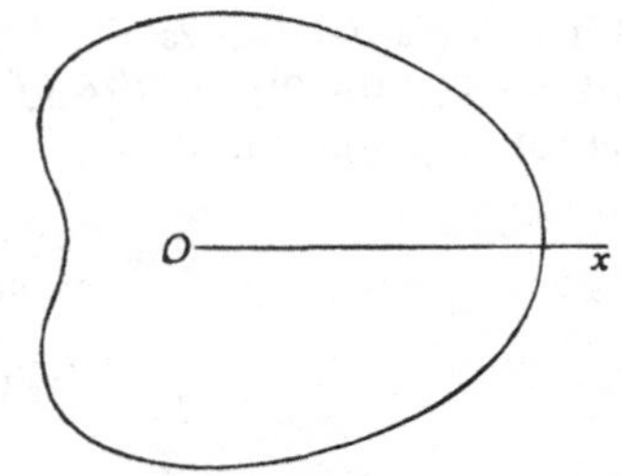

Fig. 41.

The area is

$$\begin{aligned}
&\frac{1}{2}\int_0^{2\pi} r^2\,d\theta \\
&= \frac{1}{2}\int_0^{2\pi}\left(1 + \cos\theta + \frac{1}{4}\cos^2\theta\right)d\theta \\
&= \frac{1}{2}\int_0^{2\pi}\left\{1 + \cos\theta + \frac{1}{8}(1 + \cos 2\theta)\right\}d\theta \\
&= \frac{1}{2}\int_0^{2\pi}\left\{\frac{9}{8} + \cos\theta + \frac{1}{8}\cos 2\theta\right\}d\theta \\
&= \frac{1}{2}\left[\frac{9}{8}\theta + \sin\theta + \frac{1}{16}\sin 2\theta\right]_0^{2\pi} \\
&= \frac{1}{2}\left[\left(\frac{9}{8}.2\pi\right) - 0\right] \\
&= \frac{9}{8}\pi.
\end{aligned}$$

EXAMPLES I

Find the area of each of the following curves:

1. The circle $r = 2\cos\theta$.
2. The cardioid $r = 1 + \cos\theta$.
3. The limaçon $r = 3 - 2\cos\theta$.

4. Centre of gravity, or centroid. If a number of particles $P_1, P_2, P_3, \ldots$, of masses $m_1, m_2, m_3, \ldots$, are situated at points $(x_1, y_1), (x_2, y_2), (x_3, y_3)$ in a plane (Fig. 42), their *centre of gravity* (more accurately, *centre of mass*) or *centroid* is defined to be the point $G(\xi, \eta)$, where

$$\xi = \frac{m_1 x_1 + m_2 x_2 + m_3 x_3 + \ldots}{m_1 + m_2 + m_3 + \ldots},$$

$$\eta = \frac{m_1 y_1 + m_2 y_2 + m_3 y_3 + \ldots}{m_1 + m_2 + m_3 + \ldots}.$$

Writing $M \equiv m_1 + m_2 + m_3 + \ldots$ for the total mass, we may express these relations in the form

$$M\xi = \Sigma m_k x_k, \quad M\eta = \Sigma m_k y_k,$$

where the summations extend over all the particles.

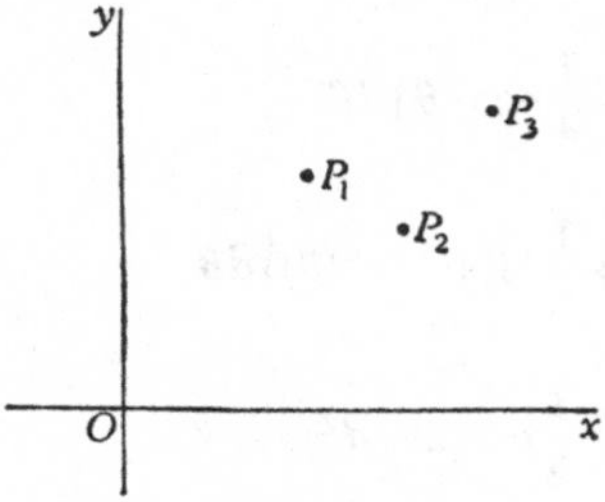

Fig. 42.

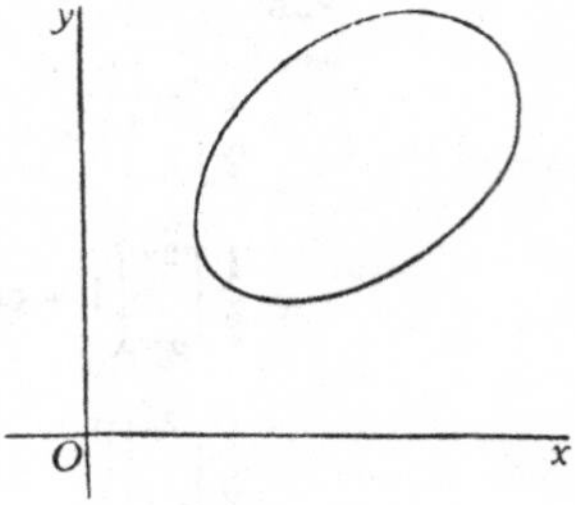

Fig. 43.

Our purpose is to extend this definition to a lamina, such as that shown in the diagram (Fig. 43), bounded by a closed curve. It is assumed that the lamina is made of uniform material. For convenience, we have placed it in the first quadrant, but that restriction is not necessary.

If the area is divided, in any way, into a large number of elements, so that a typical element surrounds the point (x_k, y_k) and has area δA_k, then, by obvious extension, the centre of gravity is the point $G(\xi, \eta)$, where

$$\xi = \frac{x_1\,\delta A_1 + x_2\,\delta A_2 + x_3\,\delta A_3 + \ldots}{\delta A_1 + \delta A_2 + \delta A_3 + \ldots},$$

$$\eta = \frac{y_1\,\delta A_1 + y_2\,\delta A_2 + y_3\,\delta A_3 + \ldots}{\delta A_1 + \delta A_2 + \delta A_3 + \ldots}.$$

The sum $\delta A_1 + \delta A_2 + \delta A_3 + \ldots$ is equal to A, the area enclosed by the curve. The remaining problem is to find an expression for the sums

$$\Sigma x_k \, \delta A_k, \quad \Sigma y_k \, \delta A_k$$

as the number of elements of area is increased indefinitely while their sizes decrease indefinitely.

We begin with the coordinate ξ.

As a first step, consider the area 'under' the curve

$$y = f(x)$$

between the ordinates $x = a$, $x = b$, where $f(x)$ is a single-valued function of x (Fig. 44). Divide the interval at the points $A \equiv M_0, M_1, M_2, \ldots, M_{n-1}, M_n \equiv B$ with x-coordinates

$$a \equiv x_0, x_1, x_2, \ldots, x_{n-1}, x_n \equiv b,$$

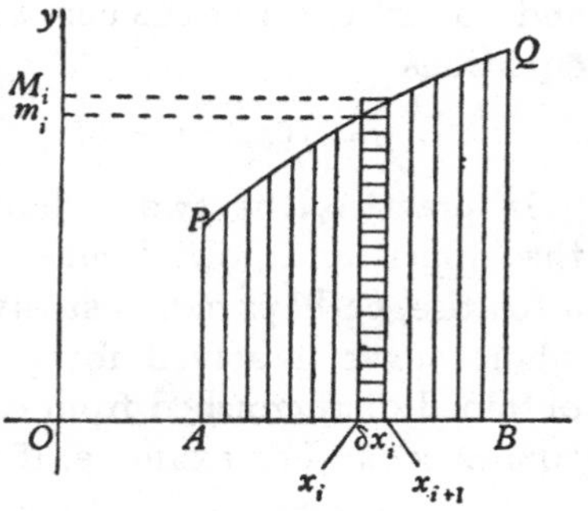

Fig. 44.

and draw the lines through these points parallel to the y-axis. (Compare the introduction to area on p. 77.)

Confining our attention to a typical filament whose base, joining the points x_i, x_{i+1} is of length $\delta x_i \equiv x_{i+1} - x_i$, let us divide it into rectangles by lines parallel to Ox and take these rectangles as the elements δA_k of the definition. Denote the least and greatest values of $f(x)$ in the interval by m_i, M_i respectively, and extend the rectangles to the height M_i. Then the contribution from the filament to the sum $\Sigma x_k \, \delta A_k$ lies between $x_i(m_i \, \delta x_i)$ and $x_{i+1}(M_i \, \delta x_i)$, so that

$$\sum_{i=0}^{n-1} (x_i m_i) \, \delta x_i \leqslant \Sigma x_k \delta A_k \leqslant \sum_{i=0}^{n-1} (x_{i+1} M_i) \, \delta x_i.$$

In the limit, for ordinary curves, the two outer sums have the same value, namely,

$$\int_a^b x f(x) \, dx,$$

and so the formula for ξ is

$$\xi = \frac{\int_a^b x f(x) \, dx}{\int_a^b f(x) \, dx}.$$

Proceeding now to the closed curve with which we began, suppose that it lies between ordinates

$$x = a, \quad x = b$$

which it meets at P, Q respectively (Fig. 45). Suppose, too, that the 'upper' and 'lower' parts of the curve have the equations

$$y = f(x), \quad y = g(x).$$

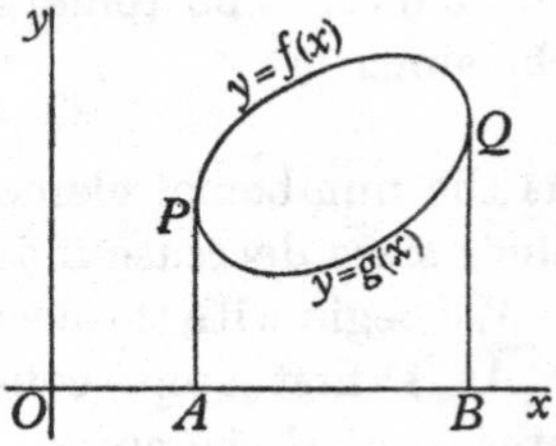

Fig. 45.

In practice, the two different forms for y may arise because the 'upper' and 'lower' curves are quite different; but they may also arise, perhaps more usually, because when the equation of the whole curve is solved for y in terms of x, two functions are obtained distinguished from each other by the sign attached to a square root. For example, if the bounding curve is the circle

$$(x-2)^2 + (y-2)^2 = 1,$$

then $$y - 2 = \pm \sqrt{\{1-(x-2)^2\}} = \pm \sqrt{(4x - x^2 - 3)}.$$

Thus $$f(x) = 2 + \sqrt{(4x - x^2 - 3)},$$

$$g(x) = 2 - \sqrt{(4x - x^2 - 3)}.]$$

Then the contribution to $\Sigma x_k \delta A_k$ from the area enclosed by the curve is equivalent to that from the area under $y = f(x)$ LESS that from the area under $y = g(x)$. Thus, if A is the area enclosed by the given curve,

$$A\xi = \int_a^b x f(x)\,dx - \int_a^b x\,g(x)\,dx.$$

This formula may be expressed more concisely. Write

$$[y]$$

to denote the difference between the two values of y corresponding to x, so that

$$[y] = f(x) - g(x).$$

Then $$\xi = \frac{\int_a^b x[y]\,dx}{A}.$$

In the same way, and with similar notation,

$$\eta = \frac{\int_c^d y[x]\,dy}{A}.$$

ALTERNATIVE EXPRESSION. *Another formula for* η *for a lamina defined by the area 'under' a curve* $y = f(x)$.

Referring to the diagram (Fig. 44) for the area $ABQP$, consider the contribution from the filament on δx_i towards the sum $\Sigma y_k \delta A_k$ required to calculate η. If the points of division of the filament are at heights

$$0 \equiv y_0, y_1, y_2, \ldots,$$

then a typical contribution is

$$y'_j(\delta y_j \, \delta x_i),$$

where
$$\delta y_j = y_{j+1} - y_j$$

and y'_j lies between y_j and y_{j+1}.

Keeping δx_i constant, let the number of subdivisions of the filament be increased indefinitely while their sizes decrease indefinitely. Then the contribution from the filament lies between

$$\int_0^{m_i} y\,dy\,\delta x_i, \quad \int_0^{M_i} y\,dy\,\delta x_i,$$

or
$$\tfrac{1}{2}m_i^2\,\delta x_i, \quad \tfrac{1}{2}M_i^2\,\delta x_i.$$

Summing now for all δx_i and proceeding to the limit in the usual way, we obtain the integral

$$\frac{1}{2}\int_a^b \{f(x)\}^2\,dx,$$

so that
$$\eta = \frac{\dfrac{1}{2}\displaystyle\int_a^b \{f(x)\}^2\,dx}{\displaystyle\int_a^b f(x)\,dx},$$

or
$$\eta = \frac{\dfrac{1}{2}\displaystyle\int_a^b y^2\,dx}{\displaystyle\int_a^b y\,dx}.$$

For an oval curve not meeting the x-axis, the corresponding formula is

$$\frac{\dfrac{1}{2}\displaystyle\int_a^b [y^2]\,dx}{\displaystyle\int_a^b [y]\,dx},$$

where
$$[y^2] = \{f(x)\}^2 - \{g(x)\}^2.$$

ILLUSTRATION 4. *To find the centre of gravity (centroid) of the area in the first quadrant bounded by the axes and the curve*

$$y = 2+x-x^2.$$

The range for x is 0, 2. The area is given by the formula

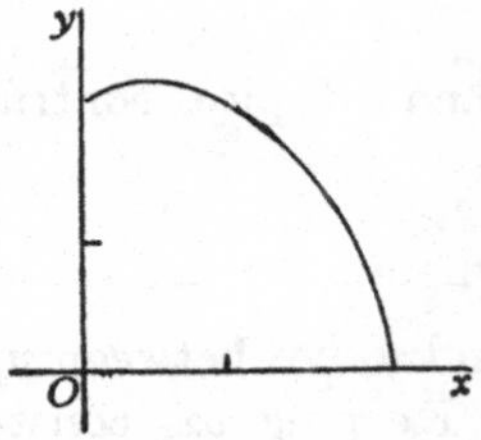

Fig. 46.

$$\int_0^2 y\,dx = \int_0^2 (2+x-x^2)\,dx$$

$$= \left[2x+\frac{1}{2}x^2-\frac{1}{3}x^3\right]_0^2$$

$$= \frac{10}{3}.$$

Then
$$\xi\int_0^2 y\,dx = \int_0^2 xy\,dx$$

$$= \int_0^2 (2x+x^2-x^3)\,dx$$

$$= \left[x^2+\frac{1}{3}x^3-\frac{1}{4}x^4\right]_0^2$$

$$= \frac{8}{3},$$

so that
$$\xi = \frac{8}{3}\Big/\frac{10}{3} = \frac{4}{5}.$$

Also
$$\eta\int_0^2 y\,dx = \frac{1}{2}\int_0^2 y^2\,dx$$

$$= \frac{1}{2}\int_0^2 (4+4x-3x^2-2x^3+x^4)\,dx$$

$$= \frac{1}{2}\left[4x+2x^2-x^3-\frac{1}{2}x^4+\frac{1}{5}x^5\right]_0^2$$

$$= \frac{16}{5},$$

so that
$$\eta = \frac{16}{5}\Big/\frac{10}{3} = \frac{24}{25}.$$

5. The moment of inertia of a lamina.

(i) THE MOMENT OF INERTIA ABOUT A LINE. If a number of particles $P_1, P_2, P_3, \ldots$, of masses $m_1, m_2, m_3, \ldots$, lie in a plane (see Fig. 42, p. 118), their *moment of inertia* about a line l in the plane is defined to be the magnitude I given by the formula

$$I = m_1 p_1^2 + m_2 p_2^2 + m_3 p_3^2 + \ldots,$$

where $p_1, p_2, p_3, \ldots$ are the perpendicular distances of $P_1, P_2, P_3, \ldots$ from l.

Whenever possible, the line l is taken to be one of the axes of coordinates. Thus, if I_y is the moment of inertia about Oy, then

$$I_y = m_1 x_1^2 + m_2 x_2^2 + m_3 x_3^2 + \ldots,$$

where $(x_1, y_1), (x_2, y_2), \ldots$ are the coordinates of $P_1, P_2, \ldots$.

Similarly,
$$I_x = m_1 y_1^2 + m_2 y_2^2 + m_3 y_3^2 + \ldots.$$

The definition can be extended to a lamina bounded by a closed curve (see Fig. 43, p. 118). The reasoning is exactly analogous to that just given (§ 4) in calculating centres of gravity, and need not be repeated. With the notation explained there, we have

$$I_y = \int_a^b x^2[y]\,dx,$$

$$I_x = \int_c^d y^2[x]\,dy.$$

(ii) THE MOMENT OF INERTIA ABOUT A POINT. The definition of the moment of inertia of a plane system of particles $P_1, P_2, \ldots$, of masses $m_1, m_2, \ldots$, about a point O in the plane is very similar, namely,

$$I_0 = m_1 OP_1^2 + m_2 OP_2^2 + \ldots.$$

If the polar coordinates of $P_1, P_2, \ldots$ are $(r_1, \theta_1), (r_2, \theta_2), \ldots$ when O is pole, then

$$I_0 = m_1 r_1^2 + m_2 r_2^2 + \ldots.$$

Alternatively, if the Cartesian coordinates of $P_1, P_2, \ldots$ are $(x_1, y_1), (x_2, y_2), \ldots$, with O as origin, then

$$I_0 = m_1(x_1^2 + y_1^2) + m_2(x_2^2 + y_2^2) + \ldots.$$

It follows that
$$I_0 = I_y + I_x,$$

so that *the moment of inertia of a plane system of particles about a point O of the plane is the sum of the moments of inertia of the* *system about two perpendicular lines lying in the plane and passing through O.*

The extension to a plane lamina is immediate.

ILLUSTRATION 5. *To find the moment of inertia about the y-axis of a uniform lamina of density* ρ *bounded by those parts of the x-axis and the curve* $y = \cos x$ *which lie between* $x = -\frac{1}{2}\pi$ *and* $x = \frac{1}{2}\pi$. (See Fig. 47.)

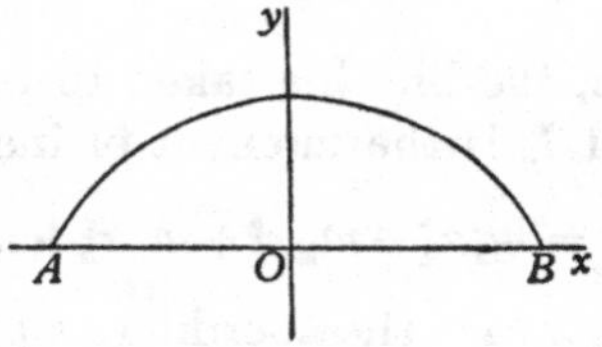

Fig. 47.

We have

$$I_y = \rho\int_{-\frac{1}{2}\pi}^{\frac{1}{2}\pi} x^2[y]\,dx$$

$$= \rho\int_{-\frac{1}{2}\pi}^{\frac{1}{2}\pi} x^2 \cos x\,dx$$

$$= \rho\int_{-\frac{1}{2}\pi}^{\frac{1}{2}\pi} x^2 \frac{d}{dx}(\sin x)\,dx$$

$$= \rho[x^2 \sin x]_{-\frac{1}{2}\pi}^{\frac{1}{2}\pi} - \rho\int_{-\frac{1}{2}\pi}^{\frac{1}{2}\pi} \sin x \,.\, 2x\,dx$$

$$= \rho\left[\left(\frac{\pi}{2}\right)^2(1) - \left(-\frac{\pi}{2}\right)^2(-1)\right] + 2\rho\int_{-\frac{1}{2}\pi}^{\frac{1}{2}\pi} x\,d(\cos x)$$

$$= \frac{\rho\pi^2}{2} + 2\rho[x\cos x]_{-\frac{1}{2}\pi}^{\frac{1}{2}\pi} - 2\rho\int_{-\frac{1}{2}\pi}^{\frac{1}{2}\pi} 1 \,.\, \cos x\,dx$$

$$= \frac{\rho\pi^2}{2} + 2\rho[0] - 2\rho[\sin x]_{-\frac{1}{2}\pi}^{\frac{1}{2}\pi}$$

$$= \frac{\rho\pi^2}{2} - 2\rho[1-(-1)]$$

$$= \frac{\rho\pi^2}{2} - 4\rho$$

$$= \frac{1}{2}\rho(\pi^2 - 8).$$

ILLUSTRATION 6. *To find the moment of inertia of a circle about its centre.*

Suppose that the equation of the circle (Fig. 48) is

$$x^2+y^2 = a^2.$$

Divide the area bounded by the circle into concentric rings, of which a typical one is enclosed by circles of radii $r, r+\delta r$. Then, by the definition of moment of inertia about a point, the moment for the ring is

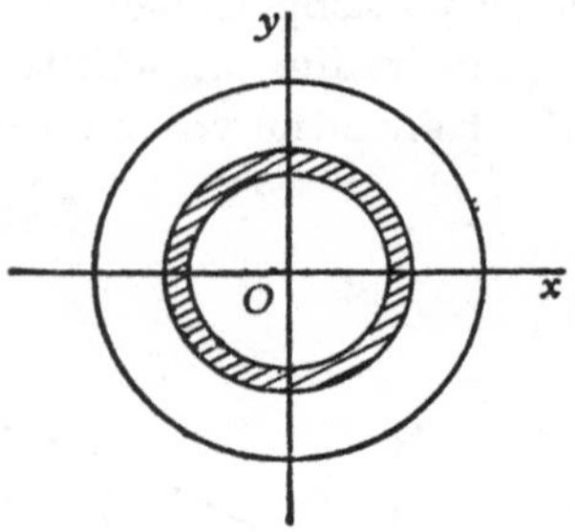

Fig. 48.

$$(2\pi r_1 \delta r) r_1^2,$$

where the number r_1 lies between r and $r+\delta r$. Adding for all the rings, and proceeding to the limit in which their number is increased indefinitely,

$$\begin{aligned} I_0 &= \int_0^a 2\pi r^3 dr \\ &= \tfrac{1}{2}\pi a^4 \\ &= \tfrac{1}{2}Aa^2, \end{aligned}$$

where A is the area of the lamina.

COROLLARY. By symmetry, $I_y = I_x$. Hence

$$I_y = I_x = \tfrac{1}{4}Aa^2.$$

6. Volume of revolution.

Let

$$y = f(x)$$

be the equation of a given curve, where $f(x)$ is a single-valued function of x, and consider the usual area $ABQP$ (Fig. 49) bounded by the curve, together with the lines $x = a, x = b, y = 0$.

Suppose that the curve does not cut the x-axis between $x = a, x = b$.

To find an expression for the volume generated by the complete revolution of this area about the axis Ox.

Suppose, as usual, that the interval a, b is divided into n parts at the points

$$a \equiv x_0, x_1, x_2, \ldots, x_{n-1}, x_n \equiv b,$$

and that

$$\delta x_i = x_{i+1} - x_i.$$

Suppose also that m_i, M_i are the least and greatest values of the function $f(x)$ in the interval δx_i. Then the volume (Fig. 50) of the corresponding element lies between that of a cylinder of height δx_i and radius m_i, and that of a cylinder of height δx_i and radius M_i. Hence the volume of revolution lies between the two sums

$$\Sigma\pi m_i^2\,\delta x_i,\quad \Sigma\pi M_i^2\,\delta x_i.$$

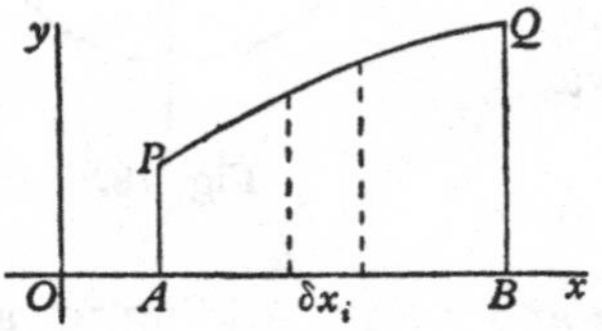

Fig. 49.

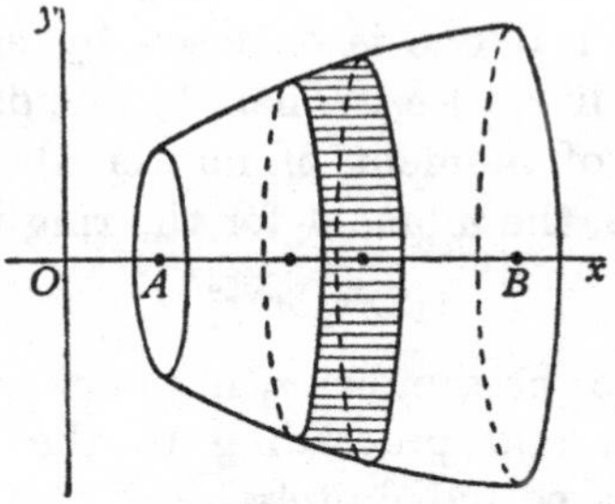

Fig. 50.

If we now allow the number of subdivisions to increase indefinitely, their lengths decreasing indefinitely, then, in ordinary cases, these two sums tend to the limit

$$\int_a^b \pi\{f(x)\}^2\,dx,$$

or

$$\int_a^b \pi y^2\,dx,$$

which is therefore the expression for the volume V.

Hence $$V = \int_a^b \pi y^2\,dx.$$

ILLUSTRATION 7. *To find the volume of a sphere of radius a.*

The sphere may be regarded as generated by the rotation of the 'upper' semi-circle of the circle

$$x^2+y^2=a^2$$

about Ox (Fig. 51).

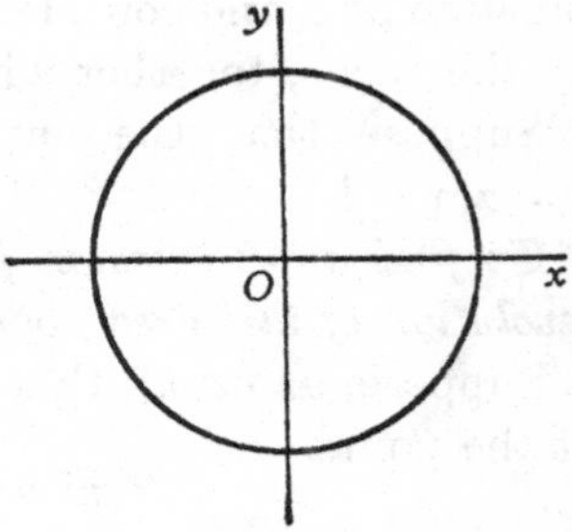

Fig. 51.

Then
$$V = \int_{-a}^{a} \pi y^2 dx$$
$$= \int_{-a}^{a} \pi(a^2 - x^2) dx$$
$$= \pi \left[a^2 x - \frac{1}{3}x^3\right]_{-a}^{a}$$
$$= \pi \left[\left(\frac{2}{3}a^3\right) - \left(-\frac{2}{3}a^3\right)\right]$$
$$= \frac{4}{3}\pi a^3.$$

7. The centre of gravity of a uniform solid of revolution. We regard it as obvious that the centre of gravity of a solid of revolution, generated as in § 6, lies on the axis Ox.

To locate the centre of gravity, we use the same principle as in § 4 (p. 118) for a lamina. In the present case, we see that, if ξ is the x-coordinate, then

$$\xi \Sigma \delta V_i = \Sigma x_i \delta V_i,$$

where δV_i is an element of volume generated by a small area of x-coordinate x_i. Dividing the area, and so the volume, into strips as in § 6, and proceeding to the limit, we obtain the formula

$$\xi V = \int_a^b \pi x y^2 dx,$$

where $V = \int_a^b \pi y^2 dx$, the volume of the solid.

ILLUSTRATION 8. *To find the centre of gravity of a uniform hemisphere.*

Regard the hemisphere as generated by the rotation about the axis Ox (Fig. 52) of that arc of the circle

$$x^2 + y^2 = a^2$$

which lies in the positive quadrant.

Its volume, as above, is

$$\frac{2}{3}\pi a^3.$$

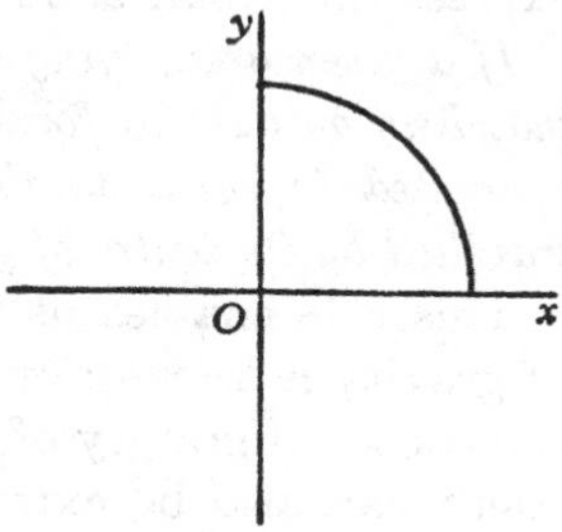

Fig. 52.

Hence $$\frac{2}{3}\pi a^3\xi = \pi\int_0^a xy^2\,dx = \pi\int_0^a x(a^2-x^2)\,dx$$
$$= \pi\left[\frac{1}{2}a^2x^2-\frac{1}{4}x^4\right]_0^a$$
$$= \frac{1}{4}\pi a^4,$$

so that $$\xi = \frac{3}{8}a.$$

8. First theorem of Pappus. Suppose that a solid of revolution is obtained by rotating the usual area $ABQP$ about the x-axis (Fig. 53). We proved (p. 126) that, if PQ is the curve
$$y=f(x),$$
the volume V so generated is given by the formula
$$V = \int_a^b \pi y^2\,dx.$$

But we also proved (p. 121) that the y-coordinate η of the centre of gravity of the plane area is given by the formula
$$\eta = \frac{\frac{1}{2}\int_a^b y^2\,dx}{A},$$
where $A \equiv \int_a^b y\,dx$ is the area $ABQP$.

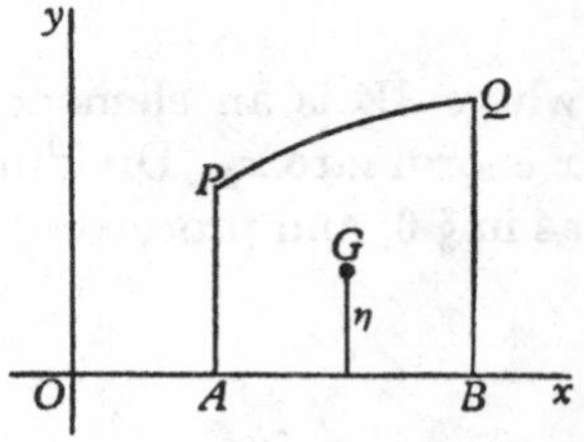

Fig. 53.

Hence $V = 2\pi A\eta$.

Since $2\pi\eta$ is the circumference of a circle of radius η, we may express this result as follows:

If a given area, lying on one side of a given line, is rotated about that line as axis to form a solid of revolution, then the volume so generated is equal to the product of the area times the distance travelled by its centre of gravity.

This rule enables us to calculate the volume when the centre of gravity is known; or, alternatively, to find the y-coordinate of the centre of gravity of the area when the volume is known. The result can also be extended easily to prove that the volume of the solid, generated by the rotation of the area bounded by a *closed curve* lying entirely 'above' the axis of rotation, is equal to the area times the distance travelled by its centre of gravity.

For example, the volume of the *anchor-ring* obtained by rotating the circle $x^2+(y-b)^2=a^2$, where $b>a$, about the x-axis is $(\pi a^2).2\pi b$, or $2\pi^2a^2b$.

ILLUSTRATION 9. *To find the centre of gravity of a semicircle of radius a.*

Rotate the semicircle about its bounding diameter. The solid generated is a sphere, whose volume is known to be $\frac{4}{3}\pi a^3$.

The area of the semicircle is $\frac{1}{2}\pi a^2$, and so the distance rotated by the centre of gravity is

$$\frac{4}{3}\pi a^3 \Big/ \frac{1}{2}\pi a^2 = \frac{8}{3}a.$$

Hence
$$2\pi\eta = \frac{8}{3}a,$$

or
$$\eta = \frac{4a}{3\pi}.$$

The centre of gravity therefore lies on the line of symmetry of the semicircle, at a distance

$$\frac{4a}{3\pi}$$

from its bounding diameter.

9. The moment of inertia of a uniform solid of revolution about its axis.

The definition given in § 5 (p. 123) for the moment of inertia of a system of particles $P_1, P_2, \ldots$, of masses $m_1, m_2, \ldots$ about a line l holds equally well in space. If $p_1, p_2, \ldots$ are the distances of $P_1, P_2, \ldots$ from l, then

$$I = m_1p_1^2+m_2p_2^2+\ldots.$$

In order to effect the summation for a solid of revolution, when l is the axis, we refer to the diagram (Fig. 50) and notation of § 6. The volume is divided into a number of circular discs, of which a typical one has radius between m_i, M_i and breadth δx_i. The moment of inertia of this solid disc about its centre is therefore between δx_i times that of a circle of radius m_i about its centre and δx_i times that of a circle of radius M_i. Hence (p. 125)

$$\sum_0^{n-1} \tfrac{1}{2}\pi m_i^4\,\delta x_i \leqslant I \leqslant \sum_0^{n-1} \tfrac{1}{2}\pi M_i^4\,\delta x_i.$$

In the limit, these two sums are equal, and so

$$I = \int_a^b \tfrac{1}{2}\pi\{f(x)\}^4 dx$$

$$= \int_a^b \tfrac{1}{2}\pi y^4 dx.$$

ILLUSTRATION 10. *To find the moment of inertia of a solid sphere about a diameter.*

Suppose that the sphere is generated by rotating the circle

$$x^2 + y^2 = a^2$$

about the diameter Ox, taken to be the given diameter. Then

$$\begin{aligned} I &= \int_{-a}^{a} \tfrac{1}{2}\pi y^4 dx \\ &= \frac{1}{2}\pi \int_{-a}^{a} (a^2 - x^2)^2 dx \\ &= \frac{1}{2}\pi \int_{-a}^{a} (a^4 - 2a^2x^2 + x^4)\, dx \\ &= \frac{1}{2}\pi \left[a^4 x - \frac{2}{3}a^2x^3 + \frac{1}{5}x^5 \right]_{-a}^{a} \\ &= \frac{1}{2}\pi \left[2a^5 - \frac{4}{3}a^5 + \frac{2}{5}a^5 \right] \\ &= \frac{1}{2}\pi \left[\frac{16a^5}{15} \right] \\ &= \frac{8}{15}\pi a^5 \\ &= \frac{2}{5} V a^2, \end{aligned}$$

where $V \equiv \frac{4}{3}\pi a^3$ is the volume of the sphere.

10.* The area of a surface of revolution.

LEMMA. *The area of a right circular cone.*

Consider a right circular cone, of *slant* height l, whose base is a circle of radius r (Fig. 54*a*). If the cone is slit down a generator and then opened up so as to lie in a plane, we obtain the sector of a circle of radius l subtended by an arc whose length, equal

* This paragraph may be postponed, if desired.

to the circumference of the base of the cone, is $2\pi r$ (Fig. 54*b*). Hence the area is the fraction $(2\pi r/2\pi l)$ of the area of a whole circle of radius l, so that

$$A = \frac{r}{l}(\pi l^2)$$
$$= \pi r l.$$

What we shall actually require is the area of a *frustum* of a cone, as indicated in the diagram (Fig. 55). If r_1, l_1 and r_2, l_2 are

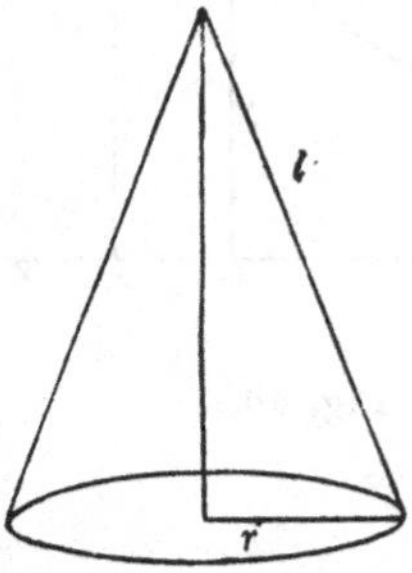

Fig. 54*a*.

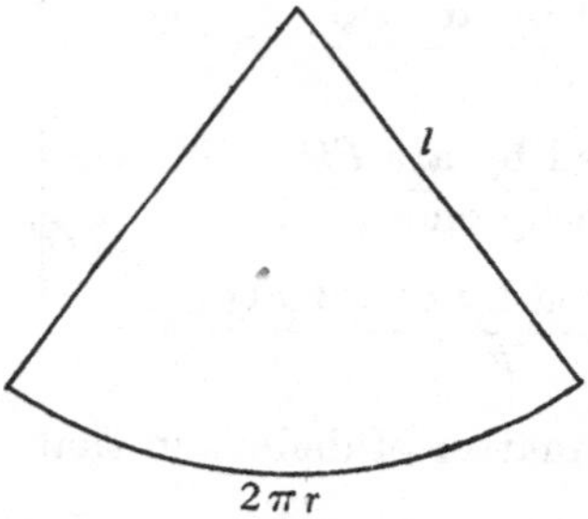

Fig. 54*b*.

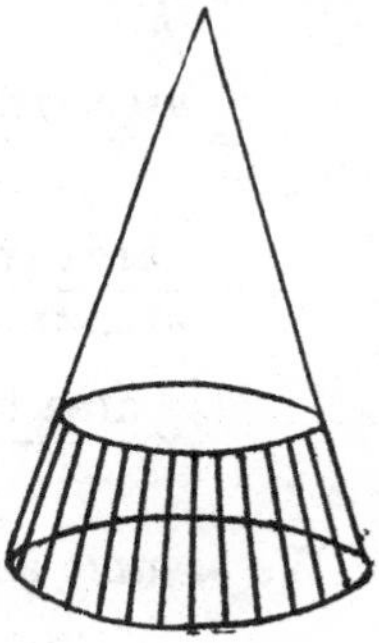

Fig. 55.

the values of r, l for the two boundaries, then the area of the frustum is

$$\pi(r_2 l_2 - r_1 l_1).$$

Now, by similarity, $$\frac{l_2}{r_2} = \frac{l_1}{r_1} = \frac{l_2 - l_1}{r_2 - r_1},$$

and so, on substituting for l_2, l_1, the area is

$$\pi(r_2^2 - r_1^2)\left(\frac{l_2 - l_1}{r_2 - r_1}\right)$$
$$= \pi(r_2 + r_1)(l_2 - l_1).$$

If we draw the 'half-way' circle through the middle of the frustum, its radius is $\frac{1}{2}(r_2 + r_1)$ and perimeter $\pi(r_2 + r_1)$. Hence the area of the frustum is

(perimeter of 'half-way' circle) × (*slant* height of the frustum).

We now proceed to our main task:

To find the area of a surface of revolution. Let $f(x)$ be a positive single-valued function of x in a certain interval (a, b), and rotate the curve

$$y = f(x)$$

through four right angles about the x-axis. We require an expression for the area of the surface generated.

Suppose that the area traced out by the arc AP is S (Fig. 56); then S is a function of x which we denote by the symbol $S(x)$.

In the plane, let $P \equiv (x, y)$, $Q \equiv (x+h, y+k)$. Then

$$\frac{S(x+h)-S(x)}{h}$$
$$= \frac{\text{area traced by arc } PQ}{h}$$
$$= \frac{\text{area traced by arc } PQ}{\text{area traced by chord } PQ} \times \frac{\text{area traced by chord } PQ}{h}$$

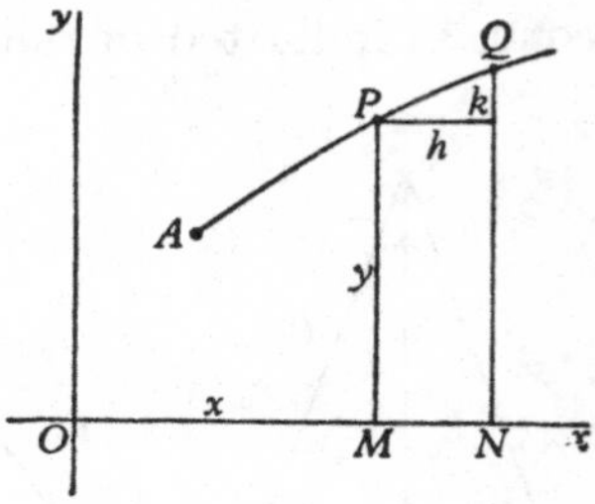

Fig. 56.

We assume as a matter of definition that

$$\lim_{h \to 0} \frac{\text{area traced by arc } PQ}{\text{area traced by chord } PQ} = 1.$$

Moreover, the area traced by the chord PQ is, in accordance with the lemma,

$$\pi(2y+k)\sqrt{(h^2+k^2)}.$$

Thus
$$S'(x) = \lim_{h \to 0} \frac{S(x+h)-S(x)}{h}$$
$$= \lim_{h \to 0} \pi(2y+k)\sqrt{\left(1+\frac{k^2}{h^2}\right)}$$
$$= \pi . 2y . \lim_{h \to 0} \sqrt{\left(1+\frac{k^2}{h^2}\right)}$$
$$= 2\pi y \sqrt{\left\{1+\left(\frac{dy}{dx}\right)^2\right\}},$$

so that,
$$S(x) = \int 2\pi y \sqrt{\left\{1+\left(\frac{dy}{dx}\right)^2\right\}}\, dx$$

between appropriate limits.

We shall see in Volume II that the integral

$$\int \sqrt{\left\{1+\left(\frac{dy}{dx}\right)^2\right\}}\, dx$$

gives the length of the arc of the curve $y = f(x)$. Denoting this length by s, we may write the expression for S in the form

$$\int 2\pi y\, ds.$$

ILLUSTRATION 11. *To find the area of that portion of a sphere of radius r which is cut off between two parallel planes at a distance h apart.*

The area is generated by rotating the arc AB of the circle

$$x^2 + y^2 = r^2$$

about the axis Ox (Fig. 57).

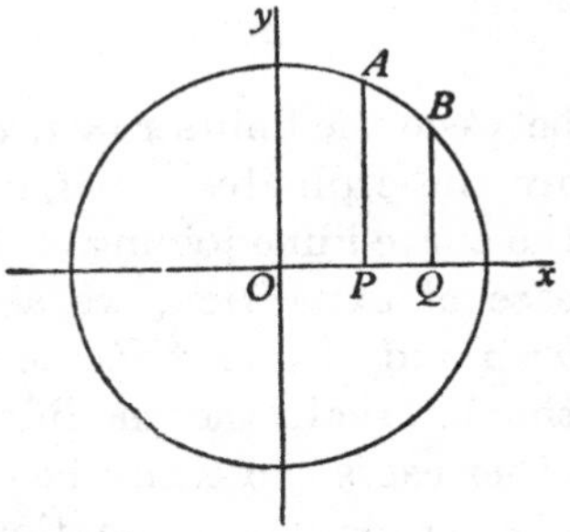

Fig. 57.

Let the x-coordinates of A, B be a, b respectively, where

$$b - a = h.$$

Differentiating the equation of the circle with respect to x, we have

$$x + y\frac{dy}{dx} = 0,$$

so that
$$1 + \left(\frac{dy}{dx}\right)^2 = 1 + \frac{x^2}{y^2} = \frac{x^2+y^2}{y^2} = \frac{r^2}{y^2},$$

and
$$\sqrt{\left\{1 + \left(\frac{dy}{dx}\right)^2\right\}} = \frac{r}{y},$$

taking positive values. Hence the area is

$$\begin{aligned} &\int_a^b 2\pi y\left(\frac{r}{y}\right) dx \\ &= \int_a^b 2\pi r\, dx = 2\pi r\Big[x\Big]_a^b \\ &= 2\pi r(b-a) \\ &= 2\pi r h. \end{aligned}$$

Note. This value depends only on r, h, and not on where the portion of the sphere is situated. It is actually equal to the area of a circular cylinder of radius r and height h.

11. Approximate integration. The evaluation of a definite integral

$$\int_a^b f(x)\,dx$$

may present considerable difficulty, but it is often easy to reach a good approximation to the result. In this and the following paragraphs we give an account of some of the methods.

We proceed by regarding the integral as the area 'under' the curve

$$y = f(x)$$

between the limits $x = a$, $x = b$. In the diagram (Fig. 58), PA, QB are the ordinates $x = a, x = b$, of lengths y_a, y_b respectively, and the curved line joining A, B is given by the equation $y = f(x)$. For ease of exposition, we assume that $b > a$ and also that $f(x)$ is positive in the interval; the modifications for other cases can easily be obtained.

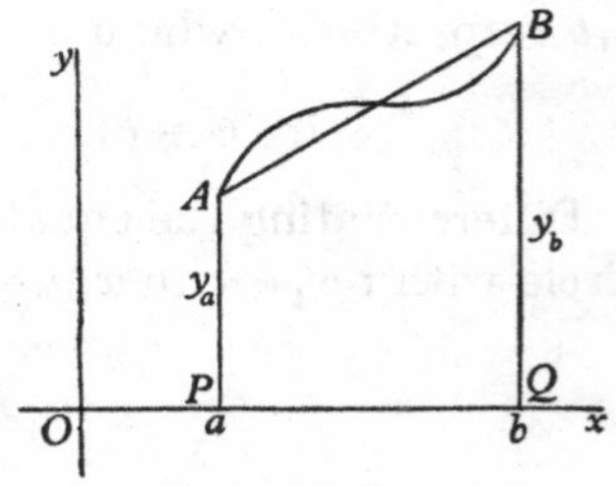

Fig. 58.

A crude approximation may be found by replacing the curve by the straight line AB, thus replacing the area under the curve by that of the trapezium $PABQ$, or

$$\tfrac{1}{2}(b-a)(y_a + y_b).$$

To improve on this, we may divide the segment PQ into say, n equal parts at points given by $x = x_1, x_2, \ldots, x_{n-1}$, and erect ordinates to the curve, of lengths $y_1, y_2, \ldots, y_{n-1}$ respectively. The curved segment AB may be replaced by the 'chain' of straight lines joining the 'tops' of these ordinates, and so we obtain n trapezia, whose areas are respectively

$$\frac{b-a}{2n}(y_a+y_1),\ \frac{b-a}{2n}(y_1+y_2),\ \frac{b-a}{2n}(y_2+y_3),\ \ldots,\ \frac{b-a}{2n}(y_{n-1}+y_b).$$

Adding, we obtain the approximation

$$\frac{b-a}{2n}\{(y_a+y_b)+2(y_1+y_2+\ldots+y_{n-1})\},$$

or $\dfrac{b-a}{2n}$ {(sum of outside ordinates) + 2(sum of inside ordinates)}.

For example, consider $\int_0^1 x^4\,dx,$

where $$f(x) \equiv x^4,$$
$$b = 1, a = 0.$$

Divide the interval into ten equal parts at the points

$$x_1 = \cdot 1,\ x_2 = \cdot 2,\ \ldots,\ x_9 = \cdot 9.$$

Then
$$y_1 = \cdot 0001 \qquad y_6 = \cdot 1296$$
$$y_2 = \cdot 0016 \qquad y_7 = \cdot 2401$$
$$y_3 = \cdot 0081 \qquad y_8 = \cdot 4096$$
$$y_4 = \cdot 0256 \qquad y_9 = \cdot 6561$$
$$y_5 = \cdot 0625$$
$$y_a = 0,\ y_b = 1.$$

The approximate value of the integral is

$$\tfrac{1}{20}\{1 + 2 \times 1\cdot 5333\} = \tfrac{1}{20}(4\cdot 0666)$$
$$= \cdot 20\dot{3}33.$$

The correct value is $\frac{1}{5}\Big[x^5\Big]_0^1 = \frac{1}{5} = \cdot 2.$

12. Simpson's rule. An approximation which is often much better than the 'trapezium' rule just given is obtained by replacing the curve $y = f(x)$ by a parabola, of the form

$$y = A + Bx + Cx^2,$$

made to pass through the end points and one intermediate point of the curve.

[The 'trapezium' rule of § 11, is, of course, equivalent to replacing the curve $y = f(x)$ by the straight line

$$y = A + Bx$$

through the end points.]

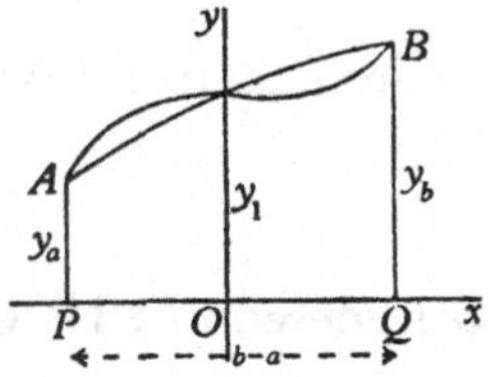

Fig. 59.

For ease of calculation, take fresh axes with the origin at the middle point of PQ, as in the diagram (Fig. 59). Let the ordinates through P, Q, O be of lengths y_a, y_b, y_1 respectively. The parabola is to pass through the points $(\frac{1}{2}(b-a), y_b)$, $(0, y_1)$, $(-\frac{1}{2}(b-a), y_a)$.

Inserting the coordinates of the three 'guide' points in the equation

$$y = A + Bx + Cx^2$$

of the parabola, and writing $h \equiv \frac{1}{2}(b-a)$, we have

$$y_b = A + Bh + Ch^2,$$
$$y_1 = A,$$
$$y_a = A - Bh + Ch^2,$$

so that

$$A = y_1,$$
$$B = \frac{1}{2h}(y_b - y_a),$$
$$C = \frac{1}{2h^2}(y_b + y_a - 2y_1).$$

Now the area under the parabola (and this area is not affected by the simplified choice of axes) is

$$\int_{-h}^{h} (A + Bx + Cx^2)\,dx$$
$$= \left[Ax + \tfrac{1}{2}Bx^2 + \tfrac{1}{3}Cx^3\right]_{-h}^{h}$$
$$= 2Ah + \tfrac{2}{3}Ch^3$$
$$= 2hy_1 + \tfrac{1}{3}h(y_b + y_a - 2y_1)$$
$$= \tfrac{1}{3}h(y_b + y_a + 4y_1)$$
$$= \tfrac{1}{6}(b-a)\{(\text{sum of outside ordinates}) + 4\,(\text{middle ordinate})\}.$$

For example, taking the integral

$$\int_0^1 x^4\,dx$$

considered in § 11, we have

$$y_a = 0, \quad y_b = 1, \quad y_1 = \tfrac{1}{16},$$

so that the approximate value is

$$\tfrac{1}{6}\{(1) + 4(\tfrac{1}{16})\} = \tfrac{1}{6}(1\tfrac{1}{4}) = \tfrac{5}{24}$$
$$= \cdot 20833,$$

which is very good agreement for so simple a calculation.

Finally, we obtain SIMPSON'S RULE as follows:

Divide the interval (a, b) into $2n$ equal parts at the points

$$x_1, x_2, \ldots, x_{2n-1},$$

and let the corresponding ordinates be

$$y_1, y_2, \ldots, y_{2n-1}.$$

Apply the preceding formula to the interval (a, x_2), giving the contribution

$$I_1 = \frac{b-a}{6n}(y_a + y_2 + 4y_1);$$

then to the interval (x_2, x_4), giving

$$I_2 = \frac{b-a}{6n}(y_2 + y_4 + 4y_3);$$

then to the interval (x_4, x_6), giving

$$I_3 = \frac{b-a}{6n}(y_4 + y_6 + 4y_5);$$

and so on, up to

$$I_n = \frac{b-a}{6n}(y_{2n-2} + y_b + 4y_{2n-1}).$$

Adding, we obtain the approximation known as *Simpson's rule for 2n divisions*:

$$\frac{b-a}{6n}\{(\text{sum of outside ordinates}) + 2(\text{sum of even ordinates}) + 4(\text{sum of odd ordinates})\}.$$

Applying this rule to $\int_0^1 x^4 dx$ with ten divisions, we have the approximation

$$\tfrac{1}{30}\{1 + 2 \times \cdot 5664 + 4 \times \cdot 9669\}$$
$$= \tfrac{1}{30}(1 + 1\cdot 1328 + 3\cdot 8676) = \tfrac{1}{30}(6\cdot 0004)$$
$$= \cdot 200013,$$

which is noticeably better than the corresponding 'trapezium' approximation (p. 135).

EXAMPLES II

Obtain approximations to the following integrals by dividing the interval of integration into ten equal parts and using (*a*) the trapezium rule, (*b*) Simpson's rule.

1. $\int_0^1 x^2\,dx.$ 2. $\int_2^4 x^3\,dx.$ 3. $\int_0^2 (x+1)^2\,dx.$

4. $\int_{-1}^1 (x^3+x)\,dx.$ 5. $\int_{-1}^1 (x^3+x^2)\,dx.$ 6. $\int_0^1 (x^3-x)\,dx.$

7. $\int_0^5 (x^4+x)\,dx.$ 8. $\int_0^1 \frac{dx}{1+x}.$ 9. $\int_1^2 \frac{dx}{1+x^2}.$

13. Mean values. It is a matter of ordinary language that the mean (or average) value of the seven numbers

$$1, 3, 8, 7, 11, 5, 4$$

is

$$\frac{1+3+8+7+11+5+4}{7}=\frac{39}{7},$$

and that the mean value of the thirteen numbers

$$2, 2, 2, \quad 3, 3, 3, 3, 3, \quad 7, \quad 9, 9, 9, 9,$$

where 2 occurs three times, 3 five times, 7 once and 9 four times, is

$$\frac{3(2)+5(3)+7+4(9)}{3+5+1+4}=\frac{6+15+7+36}{13}$$

$$=\frac{64}{13}.$$

If the numbers $y_1, y_2, \ldots, y_k$ appear $n_1, n_2, \ldots, n_k$ times respectively, the mean value is

$$\frac{n_1y_1+n_2y_2+\ldots+n_ky_k}{n_1+n_2+\ldots+n_k}.$$

If a plate is divided into k pieces, of uniform densities

$$w_1, w_2, \ldots, w_k$$

and areas

$$A_1, A_2, \ldots, A_k$$

respectively, the mean density is similarly

$$\frac{A_1w_1+A_2w_2+\ldots+A_kw_k}{A_1+A_2+\ldots+A_k}.$$

If a man walks for $\frac{1}{2}$ hour at a speed of 4 m.p.h., for $\frac{1}{4}$ hour at 5 m.p.h., rests for 20 mins., and then walks for $1\frac{1}{2}$ hours at $3\frac{1}{2}$ m.p.h., his mean speed for the whole time is, in the same way,

$$\frac{\frac{1}{2}(4)+\frac{1}{4}(5)+\frac{1}{3}(0)+\frac{3}{2}(\frac{7}{2})}{\frac{1}{2}+\frac{1}{4}+\frac{1}{3}+\frac{3}{2}} = \frac{2+\frac{5}{4}+0+\frac{21}{4}}{2\frac{7}{12}} = \frac{8\frac{1}{2}}{2\frac{7}{12}} = \frac{102}{31}$$

$$= 3\tfrac{9}{31} \text{ m.p.h.}$$

The last example serves to warn us of a danger: the distance travelled is 2 miles at 4 m.p.h., $1\frac{1}{4}$ miles at 5 m.p.h., rest for 20 mins., and then $5\frac{1}{4}$ miles at $3\frac{1}{2}$ m.p.h. The mean speed for the whole *distance* is

$$\frac{2(4)+\frac{5}{4}(5)+0(0)+\frac{21}{4}(\frac{7}{2})}{2+\frac{5}{4}+0+\frac{21}{4}} = \frac{8+\frac{25}{4}+\frac{147}{8}}{8\frac{1}{2}} = \frac{261}{68}$$

$$= 3\tfrac{57}{68} \text{ m.p.h.}$$

In other words, the mean speed with respect to time taken is NOT the same as the mean speed with respect to distance covered. Where there is any possibility of doubt, *the magnitude with respect to which a mean is calculated must be clearly stated.*

These examples illustrate the evaluation of a mean for discrete numbers, and we assume that the ideas are familiar. Our problem is to extend the conception to the mean value of a function of a continuous variable. In order to do this, we make a *definition* of the term 'mean', choosing that definition to fit in with the more elementary considerations.

DEFINITION. *The* MEAN VALUE *of the function $f(u)$, defined for all values of u in the interval (a,b), is the quotient*

$$\frac{\int_a^b f(u)\,du}{\int_a^b du},$$

or

$$\frac{1}{(b-a)}\int_a^b f(u)\,du.$$

$$\left[\text{Compare } \frac{\Sigma f(u_i)\,\delta u_i}{\Sigma\,\delta u_i}\right]$$

ILLUSTRATION 12. *To find the mean density of a straight rod of length 2a, given that the density at a distance k from the middle point is* $\rho(a+k)^2$.

Take the middle point O of the rod as the origin for a coordinate x; then $k = +x$ when x is positive and $-x$ when x is negative. The interval of integration must therefore be divided into the two parts $(-a, 0)$, $(0, a)$. Then the mean density is

$$\frac{1}{2a}\left\{\int_{-a}^{0}\rho(a-x)^2\,dx+\int_{0}^{a}\rho(a+x)^2\,dx\right\}$$

$$=\frac{1}{2a}\left\{\left[-\tfrac{1}{3}\rho(a-x)^3\right]_{-a}^{0}+\left[\tfrac{1}{3}\rho(a+x)^3\right]_{0}^{a}\right\}$$

$$=\frac{1}{2a}\{(-\tfrac{1}{3}\rho a^3+\tfrac{1}{3}\rho\,.\,8a^3)+(\tfrac{1}{3}\rho\,.\,8a^3-\tfrac{1}{3}\rho a^3)\}$$

$$=\frac{1}{2a}\left(\frac{14}{3}\rho a^3\right)$$

$$=\frac{7}{3}\rho a^2.$$

APPLICATION TO A VOLUME OF REVOLUTION.

To find the mean density of a solid of revolution whose axis is vertical, given that the density at height x above its base is

$$G(x)$$

(a function of x only), the height of the solid being h.

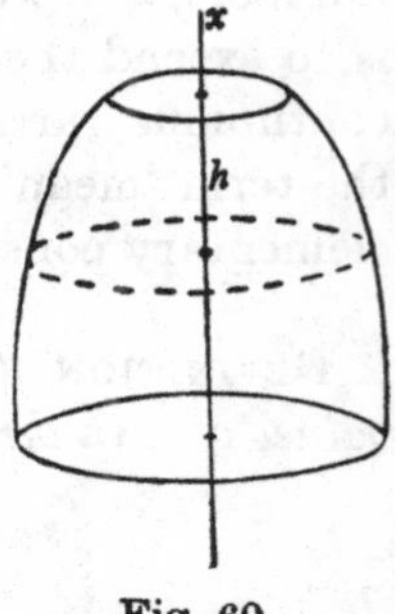

Fig. 60.

Suppose that the radius of the 'slice' (Fig. 60) at height x is r, where

$$r = f(x).$$

The volume between the base and that slice is (p. 126)

$$v \equiv \int_0^x \pi r^2\,dx,$$

so that (p. 87) $$\frac{dv}{dx} = \pi r^2.$$

The mean density with respect to volume is, by definition,

$$\frac{\int_0^V G\,dv}{\int_0^V dv},$$

where V is the total volume of the solid and where G is the function $G(x)$ regarded as a function of v in virtue of the relation connecting the two variables v, x.

Moreover, since $\frac{dv}{dx} = \pi r^2$, these integrals can be expressed in terms of x in the form

$$\frac{\int_0^h G(x)\,\pi r^2\,dx}{\int_0^h \pi r^2\,dx},$$

where $$r = f(x),$$

a given function of x.

Hence the mean density can be calculated.

ILLUSTRATION 13. *To find the mean density of a sphere of radius a, given that the density at a distance x from a given diametral plane is Ax^2.*

By the formula just obtained, with obvious modification, the mean density is

$$\frac{\int_{-a}^{a} Ax^2\,\pi r^2\,dx}{\int_{-a}^{a} \pi r^2\,dx},$$

where, for a sphere, $$r^2 + x^2 = a^2.$$

Now $$\int_{-a}^{a} \pi r^2\,dx = \int_{-a}^{a} \pi(a^2 - x^2)\,dx = \pi\Big[a^2x - \tfrac{1}{3}x^3\Big]_{-a}^{a}$$
$$= \tfrac{4}{3}\pi a^3,$$

as is familiar.

Also $$\int_{-a}^{a} \pi Ax^2 r^2\,dx = \int_{-a}^{a} \pi Ax^2(a^2 - x^2)\,dx$$
$$= \pi A\Big[\tfrac{1}{3}a^2x^3 - \tfrac{1}{5}x^5\Big]_{-a}^{a}$$
$$= \tfrac{4}{15}\pi Aa^5.$$

Hence the mean density is

$$\tfrac{1}{5}Aa^2.$$

APPLICATION TO A SURFACE OF REVOLUTION.* *To find the mean density of a surface of revolution whose axis is vertical, given that the density at height x above its base is*

$$G(x)$$

(a function of x only), the height of the surface being h.

Suppose that the radius of the 'ring' at height x is r, where

$$r = f(x).$$

The area of surface between the base and that slice is (p. 132)

$$S = \int_0^x 2\pi r \sqrt{\left\{1 + \left(\frac{dr}{dx}\right)^2\right\}}\, dx,$$

so that (p. 87) $$\frac{dS}{dx} = 2\pi r \sqrt{\left\{1 + \left(\frac{dr}{dx}\right)^2\right\}}.$$

The mean density with respect to area is, by definition,

$$\frac{\int_0^A G dS}{\int_0^A dS},$$

where A is the total area of the surface, and where G is the function $G(x)$ regarded as a function of S in virtue of the relation connecting the two variables S, x.

Since $$\frac{dS}{dx} = 2\pi r \sqrt{\left\{1 + \left(\frac{dr}{dx}\right)^2\right\}},$$

these integrals can be expressed in terms of x in the form

$$\frac{\int_0^h G(x) \,.\, 2\pi r \sqrt{\left\{1 + \left(\frac{dr}{dx}\right)^2\right\}}\, dx}{\int_0^h 2\pi r \sqrt{\left\{1 + \left(\frac{dr}{dx}\right)^2\right\}}\, dx},$$

where $$r = f(x),$$

a given function of x.

Hence the mean density can be calculated.

* This application may be postponed, if desired.

ILLUSTRATION 14. *A bowl of depth a is formed by rotating that part of the parabola*

$$y^2 = 4ax$$

for which $0 \leqslant x \leqslant a$ *about the x-axis. The density of the 'ring' at position x is* $a+x$. *To find the mean density of the bowl.* (See Fig. 61.)

Adapting the formula given above, the mean density is

$$\frac{\int_0^a (a+x).2\pi y\sqrt{\left\{1+\left(\frac{dy}{dx}\right)^2\right\}}\,dx}{\int_0^a 2\pi y\sqrt{\left\{1+\left(\frac{dy}{dx}\right)^2\right\}}\,dx}.$$

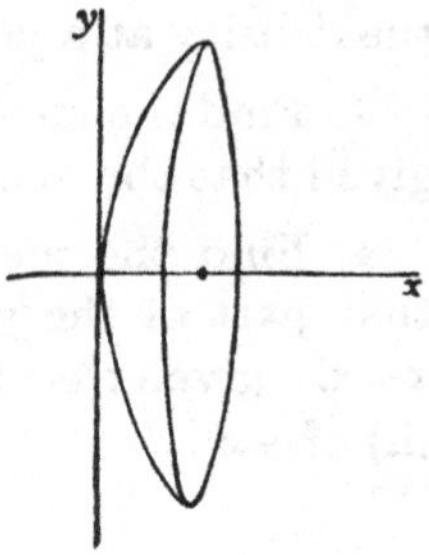

Fig. 61.

Now $\qquad y^2 = 4ax,$

so that $\qquad 2y\frac{dy}{dx} = 4a,$

and

$$1+\left(\frac{dy}{dx}\right)^2 = 1+\frac{4a^2}{y^2} = 1+\frac{4a^2}{4ax}$$

$$= \frac{(a+x)}{x}.$$

Hence the mean density is

$$\frac{\int_0^a (a+x).2\pi\sqrt{(4ax)}.\sqrt{\left(\frac{a+x}{x}\right)}\,dx}{\int_0^a 2\pi\sqrt{(4ax)}.\sqrt{\left(\frac{a+x}{x}\right)}\,dx},$$

The numerator is

$$4\pi\sqrt{a}\int_0^a (a+x)^{\frac{3}{2}}dx = 4\pi\sqrt{a}.\tfrac{2}{5}\Big[(a+x)^{\frac{5}{2}}\Big]_0^a$$

$$= \tfrac{8}{5}\pi\sqrt{a}\{(2a)^{\frac{5}{2}} - (a)^{\frac{5}{2}}\} = \tfrac{8}{5}\pi a^3(4\sqrt{2}-1)$$

The denominator is

$$4\pi\sqrt{a}\int_0^a (a+x)^{\frac{1}{2}}dx = 4\pi\sqrt{a}.\tfrac{2}{3}\Big[(a+x)^{\frac{3}{2}}\Big]_0^a$$

$$= \tfrac{8}{3}\pi\sqrt{a}\{(2a)^{\frac{3}{2}} - (a)^{\frac{3}{2}}\} = \tfrac{8}{3}\pi a^2(2\sqrt{2}-1).$$

The mean density is thus $\quad \dfrac{3}{5}\dfrac{(4\sqrt{2}-1)}{(2\sqrt{2}-1)}a,$

or, on multiplying numerator and denominator by $2\sqrt{2}+1$,

$$\frac{3(4\sqrt{2}-1)(2\sqrt{2}+1)}{5(8-1)}.a = \frac{3a}{35}(15+2\sqrt{2})$$

EXAMPLES III

1. ABC is an equilateral triangle of side $2a$, and a point P is taken on the side BC. Find the mean value of AP^2 as P varies on BC.

2. Find the mean density of a rod AB of length $2l$, given that the density at a point P is AP^2+PB^2.

3. Find the mean density of a thin hemispherical cap of radius a, given that the density at a distance x from the base is $a+x$.

4. Find the mean density of a thin bowl formed by rotating that part of the parabola $y^2 = 4ax$ for which $0 \leqslant x \leqslant a$ about the x-axis, given that the density of the ring at position x is (i) $(a+x)^2$, (ii) a^2+x^2.

REVISION EXAMPLES II

'Alternative Ordinary' Level

1. Integrate $(2x-3)^2$ and $2x^{\frac{1}{2}}$ with respect to x.

2. Evaluate the integrals

$$\int_{-1}^{1} x^2\,dx, \quad \int_{-1}^{1} x^4\,dx, \quad \int_{1}^{2} \frac{dx}{x^2}.$$

How could you show graphically, without evaluation, that the value of the first integral is greater than the value of the second?

3. Find an expression which, when differentiated with respect to x, gives

$$x^4+2+\frac{1}{x^4}.$$

Find the value of the integral

$$\int_{-1}^{1} (x+1)^2\,dx.$$

4. Integrate with respect to x:

(i) $x^2(1-x)^2$; (ii) $\sin^2 x$.

Evaluate $$\int_0^{\frac{1}{2}\pi} \sin 2x \cos x\,dx.$$

5. Integrate with respect to x:

(i) $(x+1)^3$; (ii) $\cos^2 3x$.

Evaluate $$\int_0^{\frac{1}{2}\pi} \sin^2 x \cos x\,dx$$

6. Integrate the following with respect to x:

(i) $\sin^2 x$; (ii) $\dfrac{1-x^4}{x^2}$; (iii) $\sin 2x \cos x$.

7. Calculate $$\int_a^{2a}\left(3-\frac{a^2}{x^2}+\frac{2x^3}{a^3}\right)dx.$$

8. A curve has a gradient which is given by $\dfrac{dy}{dx} = 3x-2$; also this curve cuts the y-axis at the point where $y = 4$. Find the equation of the curve, and the equation of the normal at the point $(0, 4)$.

9. Find the equation of the curve which passes through the point $(0, 1)$ and whose gradient at the point (x, y) is given by

$$\frac{dy}{dx} = 4x^3 - x.$$

Find the values of x for which y is a minimum and draw a rough sketch of the curve.

10. The gradient of a curve at the point (x, y) is $1-\dfrac{x^2}{2}$. Find the equation of the curve if it passes through the point $(2, 4)$.

Find the point of contact of the tangent which is parallel to the tangent at $(2, 4)$; also find the equations of both of these tangents.

11. Prove that the x-axis is a tangent to the curve $y = (2x-1)^2$.

Find the area bounded by this curve, the x-axis, and the line $x = 1$.

Find the area bounded by the curve, the x-axis, and the tangent at the point $(1, 1)$.

12. A curve passes through the point $(2, \frac{7}{2})$ and its gradient at the point (x, y) is $1-\dfrac{1}{x^2}$. Find the equation of the curve.

Where has y a minimum value on this curve?

13. Calculate the area bounded by the curve $y = x^2+2$ and the lines $y = x$, $x = 1$ and $x = 3$.

Calculate the area in the first quadrant bounded by the curve $y = x^2$ and the lines $x = 0$, $y = 1$ and $y = 4$.

14. A particle starts with a velocity of 2 ft. per sec. and moves along a straight line. Its acceleration in ft. per sec. per sec. after t sec. is $t+3$. Find its velocity at the end of 2 sec. and the distance travelled in the next 2 sec.

15. A particle starts from rest and moves in a straight line. Its velocity in ft. per sec. is given to be $8t-t^2$, where t is the time in seconds from the commencement of motion. How far will the particle have moved in 3 seconds?

Find also its greatest distance from the starting point, and the value of t when this distance is reached.

16. The velocity of a train starting from rest is proportional to t^2, where t is the time which has elapsed since it started. If the distance it has covered at the end of 6 seconds is 18 ft., find the velocity and the rate of acceleration at that instant.

17. A particle moves in a straight line with velocity $7t-t^2-6$ ft. per sec. at the end of t seconds. What is its acceleration when $t = 2$ and when $t = 4$?

When $t = 3$ the particle is at A; when $t = 5$ the particle is at B. Find the length of AB.

For what values of t is the particle momentarily at rest?

18. A train starts from rest and its acceleration t sec. after the start is $\frac{1}{4}(20-t)$ ft. per sec. per sec. What is its speed after 20 sec.?

Acceleration ceases at this instant and the train proceeds at this uniform speed. What is the total distance covered 30 sec. after the start from rest?

19. A body starts with velocity zero from a fixed point O and moves in a straight line; its acceleration t secs. after it leaves O is $3-t$ ft. per sec. per sec. Find the velocity of the body 4 sec. after leaving O and the distance travelled in the third second of the motion.

20. The velocity v of a particle moving in a straight line OA is given by the equation

$$\frac{d(v^2)}{dx} = -8x,$$

where x is the distance from O and $v = 3$ when $x = 2$. Find v when $x = 0$; also find the greatest positive value which x attains during the motion.

21. The velocity of a particle moving in a straight line is observed to be $4t+4t^2-t^3$ cm. per sec. at the end of t sec. Find the acceleration of the point after 4 sec., the distance of the point after 5 sec. from its position when $t = 0$, and the distance travelled in the fourth second.

22. A body starts from a point O and moves in a straight line. Its velocity at O is zero and its acceleration t sec. after leaving O is $5-\frac{2}{5}t$ ft. per sec. per sec. Find the greatest velocity attained by the body on its outward journey, and its distance from O at the instant when it begins to return towards O.

23. A particle moving in a straight line has an acceleration of $3-t$ ft. per sec. per sec. at time t sec. When $t = 1$ the particle is at rest at a point A. Find for what value of t greater than 1 the particle is again at rest and how far it is then distant from A.

24. The velocity v (in ft. per sec.) of a particle moving in a straight line is given by $v = t^2-7t+10$, where t is the number of seconds which have elapsed since the particle passed through a fixed point O on the line. Show that the particle is momentarily at rest at each of two points A and B on the line and find the length of AB.

25. The velocity v in feet per second of a point moving in a straight line is given by $v = 3t^2+2t+1$, where t is the time in seconds that has elapsed since a given instant. Find the acceleration when $t = 2$ and the distance covered between the times $t = 2$ and $t = 3$.

26. A car starts from rest with a variable acceleration, its acceleration after t seconds being $(a-3t)$ feet per second per second. If the distance covered in the first 4 seconds is 88 feet, find the value of a.

27. From the point $P(2, 4)$ on the curve $y = x^2$, PN is drawn perpendicular to the axis of x. Find the area bounded by PN, the axis of x and the curve.

Find also the x and y coordinates of the centre of gravity of this area.

28. Find the area bounded by the curve

$$y = (x+1)(x-2)^2$$

and the x-axis from $x = -1$ to $x = 2$.

Find also the x-coordinate of the centre of gravity of this area.

29. Calculate $$\int_{-1}^{1} x(x^2-1)\,dx.$$

Find the area bounded by the curve $y = x(x^2-1)$ and the x-axis (i) between $x = -1$ and $x = 0$ and (ii) between $x = 0$ and $x = 1$. Explain with the aid of a rough figure the connexion between your results and the value of the integral found in the first part.

30. Find the coordinates of the centre of gravity of the area which lies above the (positive) x-axis and below the curve $y = x^2(3-x)$.

31. Find the area bounded by the curve $y = x^2$, the axis of x and the ordinates $x = 1$ and $x = 2$.

Find the x and y coordinates of the centre of gravity of this area.

32. Calculate the coordinates of the centre of gravity of the area enclosed by the straight lines $x = 0$, $y = 0$ and the portion of the curve $y = 9-x^2$ which lies in the first quadrant.

33. Find the x-coordinate of the centre of gravity of the area bounded by the x-axis, the y-axis, and that portion of the curve

$$y = (x+1)(4-x)$$

which lies in the first quadrant.

34. Find the area included between the axis of x and the portion of the curve $y = x^2-9$ below that axis.

Find also the coordinates of the centre of gravity of this area.

35. A flat thin plate of uniform density is bounded by the two curves $y = x^2$, $y = -x^2$ and the line $x = 2$. Find its area and the coordinates of its centre of gravity.

36. Calculate the area above the x-axis bounded by the curve $y = 2x(3-x)$ and the x-axis.

Find both coordinates of the centre of gravity of this area.

37. A uniform lamina in the form of a quadrant of a circle of radius a is bounded by radii OA and OB. Find the distance of the centre of gravity of the lamina from OA and from OB.

By considering the rotation of the lamina about OA, prove that the volume of a hemisphere of radius a is $\frac{2}{3}\pi a^3$.

38. The area enclosed by the parabola $y^2 = 4x$ and the straight line $x = 4$ is rotated about the axis of x through two right angles. Find the volume of the solid so generated and the x-coordinate of its centre of gravity.

39. An area in the first quadrant is bounded by the ellipse $4x^2 + 9y^2 = 36$ and the axes of coordinates. This area is rotated through four right angles about the x-axis. Find (i) the volume generated and (ii) the x-coordinate of the centre of gravity of this volume.

40. The area bounded by the arc of the curve $y = x(3-x)$ between the points when $x = 0$ and $x = 2$, the x-axis, and the line $x = 2$, is rotated about the x-axis. Find the volume of the solid of revolution so generated, and the x-coordinate of its centre of gravity.

41. The curve $y^2 = x^2(2-x)$ cuts the x-axis at the points given by $x = 0$ and $x = 2$. The area enclosed by the x-axis and the curve between these two points is rotated through four right angles about the axis of x so as to form a solid of revolution. Find the volume of this solid and the x-coordinate of its centre of gravity.

42. Solids of revolution are generated by rotating (i) about the x-axis the area bounded by the arc of the curve $y = 2x^2$ between $(0, 0)$ and $(2, 8)$, the line $x = 2$ and the x-axis; (ii) about the y-axis the area bounded by the same arc, the line $y = 8$ and the y-axis. Calculate the volumes of the two solids so formed.

43. A cylindrical hole of radius 4 in. is cut from a sphere of radius 5 in., the axis of the cylinder coinciding with a diameter of the sphere. Prove that the volume of the remaining portion of the sphere is 36π.

44. Sketch the curve whose equation is $y = x^3$ and find the area bounded by the curve, the positive x-axis, and the straight line $x = 2$.

Find also the volume generated when this area is rotated about the x-axis through four right angles.

45. A uniform solid right circular cone is of height h and the radius of its base is r. Find the volume by the methods of the integral calculus.

Find also the distance of the centre of gravity of the cone from its vertex.

46. The radius of a sphere is 5 in. Two parallel planes are drawn at distances 2 and 3 in. respectively from the centre and 1 in. apart. Use the calculus to determine the volume of the slice of the sphere between the two planes.

[Regard the sphere as formed by the rotation of the circle $x^2+y^2 = 25$ about the x-axis.]

47. Sketch roughly the two curves

$$x^2+y^2 = 25, \quad x^2+4y^2 = 25.$$

A solid is formed by the revolution through four right angles about the x-axis of the part of the area between the two curves in which y is positive. Find the volume of the solid.

48. Find the volume of the solid of revolution generated by the rotation about the x-axis of the area bounded by the curve $y^2 = 2x$ and the line $y = \frac{1}{2}x$.

49. The ellipse
$$\frac{x^2}{a^2}+\frac{y^2}{b^2} = 1$$
is revolved about the x-axis. Find the volume of the solid so formed and the x-coordinate of the centre of gravity of its right-hand half.

50. The area between the circle $x^2+y^2 = 16$ and the ellipse $9x^2+16y^2 = 144$ is rotated about the x-axis. Calculate the volume of the solid of revolution so formed.

51. A solid of revolution is formed by rotating the portion of the curve $y = a\sin x$ between $x = 0$, $x = \frac{1}{2}\pi$ about the x-axis. Find the volume of the solid and the distance of its centre of mass from the origin.

52. Find the area included between the curve $y = x+x^2+3x^3$, the axis of x, and the ordinates at $x = 1, x = 3$. Show also that the area is exactly equal to that of a rectangle on the same base (of two units) whose height is $\frac{1}{6}(y_1+4y_2+y_3)$, where y_1, y_2, y_3 are the ordinates at $x = 1, x = 2, x = 3$ respectively.

53. Apply Simpson's rule to calculate approximately

$$\int_0^{\frac{1}{2}\pi} \sqrt{(2+\sin x)}\, dx,$$

making use of the table:

x	0	$\frac{1}{8}\pi$	$\frac{1}{4}\pi$	$\frac{3}{8}\pi$	$\frac{1}{2}\pi$
$\sqrt{(2+\sin x)}$	1·414	1·544	1·645	1·710	1·732

54. Prove that

$$\int_a^b y\, dx = \tfrac{1}{8}(b-a)\{y_b + y_a + 3(y_1 + y_2)\},$$

where y is a polynomial in x of the *third* degree, y_a, y_b are the values of y corresponding to the end points, and y_1, y_2 are the values of y corresponding to the points of trisection of the interval a, b.

Hence obtain an approximate value for

$$\int_0^{0\cdot 3} (1 - 8x^3)^{\frac{1}{2}}\, dx.$$

55. Use Simpson's rule, taking five ordinates [four divisions], to find an approximation to two decimal places to the value of the integral

$$\int_1^2 \sqrt{\left(x - \frac{1}{x}\right)}\, dx.$$

56. A river is 80 feet wide. The depth d in feet at a distance x feet from one bank is given by the following table:

x	0	10	20	30	40	50	60	70	80
d	0	4	7	9	12	15	14	8	3

Find approximately the area of the cross-section.

57. Establish Simpson's rule that, if

$$f(x) \equiv A + Bx + Cx^2 + Dx^3,$$

then

$$\int_0^1 f(x)dx = \tfrac{1}{6}\{f(0) + f(1) + 4f(\tfrac{1}{2})\}.$$

Prove also that, if

$$f(x) \equiv A + Bx + Cx^2 + Dx^3 + Ex^4,$$

then the error in still using that rule is $\frac{1}{120}E$.

58. The speed v in miles per hour of a train starting from rest is observed at intervals of one minute:

t	0	1	2	3	4	5	6	7	8
v	0	7	13	18	22	25	28	33	27

Estimate the distance covered in the eight minutes.

59. Show that constants a, b, c exist such that, if $f(x)$ is any polynomial function of degree five (or less), then

$$\int_{-2h}^{2h} f(x)\,dx = h\{af(0)+b[f(h)+f(-h)]+c[f(2h)+f(-2h)]\}.$$

Show also that, with these values of a, b, c but with $f(x) \equiv x^6$, the expression on the right is equal to $\frac{7}{6}\int_{-2h}^{2h} f(x)\,dx$.

60. Evaluate $$\int_0^{\frac{1}{3}\pi} \sin^{\frac{3}{2}} x\,dx$$

approximately by Simpson's rule, using five ordinates (i.e. at intervals of $\frac{1}{12}\pi$).

61. Evaluate $$\int_0^{\frac{1}{2}\pi} \theta \sin^2\theta\,d\theta$$

to three decimal places both by an exact method and by Simpson's rule using five ordinates.

62. If $y = a+bx+cx^2+dx^3$, prove that

$$\int_0^{3h} y\,dx = \tfrac{3}{8}h(y_0+3y_1+3y_2+y_3),$$

where y_0, y_1, y_2, y_3 on the values of y at $x = 0, h, 2h, 3h$ respectively.

Hence approximate to the value of

$$\int_0^{\frac{1}{2}\pi} (1+8\sin^2 x)^2\,dx.$$

63. Using Simpson's rule with seven ordinates, calculate the area under the curve

$$y = x^3+3$$

between the ordinates $x = 0$, $x = 6$.

ANSWERS TO EXAMPLES

CHAPTER I

Examples I:

1. (i) $x^2y - y^3 = 4$, (ii) $x^2 + y^2 - \sqrt{y} = 2$.
2. $a^2 = b^2 + c^2$, $c = \sqrt{(a^2 - b^2)}$.

Examples II:

1. 1; 1 and -1; none; 1 and 3.
2. $n\pi$; $\frac{1}{2}(2n+1)\pi$.
3. 0; 1 and -1.

Examples IV:

1. $\frac{1}{2}$.	2. -1.	3. 1.	4. 0.
5. $\frac{5}{3}$.	6. 4.	7. -2.	8. 0.

Examples V:

1. 1.	2. 1 and -1.	3. 3 and -3.
4. 1 and 2.	5. $n\pi$.	6. $\frac{1}{2}(2n+1)\pi$.
7. $\frac{1}{2}(2n+1)\pi$.	8. 1.	9. 3 and -3.
10. $\frac{1}{2}(4n+1)\pi$.	11. None.	12. $(2n+1)\pi \pm \frac{1}{3}\pi$.

Examples VI:

1. 3 at each value of x.
2. -2, 0, 2, 4.
6. $1 < x < 2$.
8. $2x - y - 1 = 0$, $4x + y + 4 = 0$, $y = 0$, $6x - y - 9 = 0$.
9. $4x - y - 1 = 0$, $y + 1 = 0$, $2x - y = 0$.

Examples VII:

9. 4, 20, 12.

Examples VIII:

1. Tangent $30x - y - 45 = 0$, normal $x + 30y - 1353 = 0$.
   Tangent $20x + y + 20 = 0$, normal $x - 20y + 402 = 0$.
   Tangent $y = 0$, normal $x = 0$.
2. Tangent $12x - y - 16 = 0$, normal $x + 12y - 98 = 0$.
   Tangent $3x - y + 2 = 0$, normal $x + 3y + 4 = 0$.
   Tangent $y = 0$, normal $x = 0$.
3. (i) Tangent $12x - y - 24 = 0$, normal $x + 12y - 2 = 0$.
   (ii) Tangent $y + 8 = 0$, normal $x = 0$.

CHAPTER II

Examples I:

1. $4x^3$.
2. $4(x+3)^3$.
3. $8(2x+3)^3$.
4. $4x^3 + 2x$.
5. $3x^2(x+1)^2(2x+1)$.
6. $2(x+1)(x+2)(2x+3)$.
7. $-x^{-2}$.
8. $-4x^{-5}$.
9. $-3(x+2)^{-4}$.
10. $-10(2x+3)^{-6}$.
11. $-21(3x-5)^{-8}$.
12. $-8(4x+3)^{-3}$.
13. $\frac{1}{2}x^{-\frac{1}{2}}$.
14. $\frac{2}{3}x^{-\frac{1}{3}}$.
15. $\frac{1}{2}x^{-\frac{1}{2}}(3x+1)$.
16. $(2x+3)^{-\frac{1}{2}}$.
17. $4(5x+7)^{-\frac{1}{5}}$.
18. $-\frac{5}{4}x^{-\frac{9}{4}}$.
19. $-\frac{1}{2}(x+7)^{-\frac{3}{2}}$.
20. $(3x+2)(2x+5)^{-\frac{1}{2}}$.
21. $(x+1)^{-2}$.
22. $6x(2x+3)^{-3}$.
23. $-\frac{1}{2}x^{-\frac{1}{2}}(20x+1)(4x-1)^{-4}$.
24. $\frac{1}{6}(x-5)(x+1)^{-\frac{1}{2}}(x-1)^{-\frac{4}{3}}$.

Examples II:

1. $2\cos 2x$.
2. $-3\sin 3x$.
3. $25\cos 5x$.
4. $\sin 2x + 2x\cos 2x$.
5. $2x\cos x - x^2\sin x$.
6. $3\cos 3x - 9x\sin 3x$.
7. $2(x+1)\sin 7x + 7(x+1)^2\cos 7x$.
8. $3\cos(3x+5)$.
9. $6(2x+1)^2$.
10. $-4(x+2)^{-5}$.
11. $\operatorname{cosec} x - x\operatorname{cosec} x\cot x$.
12. $1 + \sin x + x\cos x$.

13. $\sin 2x$.
14. $-\sin 2x$.
15. $3\sin^2 x\cos x$.
16. $-3\sin x\cos^2 x$.
17. $\sin^2 x + x\sin 2x$.
18. $2x\cos^2 x - x^2\sin 2x$.
19. $2x\cos^2 2x - 2x^2\sin 4x$.
20. $\sin^2 x + (1+x)\sin 2x$.
21. $-\sin(x - \frac{1}{4}\pi)$.
22. $\cos 2x$.
23. $\frac{1}{2}x^{-\frac{1}{2}}$.
24. $-\frac{1}{2}x^{-\frac{3}{2}}$.
25. $\frac{2}{3}x^{-\frac{1}{3}}\sin^2 x + x^{\frac{2}{3}}\sin 2x$.
26. $\frac{1}{3}\cos x(\sin x)^{-\frac{2}{3}}$.
27. $(1 + \frac{1}{2}x\cot x)(\sin x)^{\frac{1}{2}}$.
28. $2x(\sin 4x)^{\frac{1}{2}}(1 + x\cot 4x)$.
29. $-\operatorname{cosec} x\cot x$.
30. $\sec x\tan x$.
31. $\sec^2 x$.
32. $-\operatorname{cosec}^2 x$.
33. $-\dfrac{\pi}{90}\sin 2x^\circ$.
34. $\dfrac{\pi}{60}\sec^2 3x^\circ$.
35. $\sin x^\circ + \dfrac{\pi x}{180}\cos x^\circ$.
36. $\dfrac{\pi}{180}\sin 2x^\circ$.

Examples III:

1. $5x^4$, $20x^3$, $60x^2$.
2. $3x^2$, $6x$, 6.
3. 1, 0, 0.
4. $\sin x + x\cos x$, $2\cos x - x\sin x$, $-3\sin x - x\cos x$.
5. $2x\cos x - x^2\sin x$, $2\cos x - 4x\sin x - x^2\cos x$,
   $-6\sin x - 6x\cos x + x^2\sin x$.
6. $\sin^2 x + x\sin 2x$, $2\sin 2x + 2x\cos 2x$, $6\cos 2x - 4x\sin 2x$.
7. $2\cos 2x$, $-4\sin 2x$, $-8\cos 2x$.
8. $-4\sin 4x$, $-16\cos 4x$, $64\sin 4x$.
9. $\sin 2x$, $2\cos 2x$, $-4\sin 2x$.
10. $-x^{-2}$, $2x^{-3}$, $-6x^{-4}$.
11. $-2(2x-3)^{-2}$, $8(2x-3)^{-3}$, $-48(2x-3)^{-4}$.
12. $-3x^{-4}$, $12x^{-5}$, $-60x^{-6}$.

Examples IV:

1. $2\sec 2x\tan 2x$.
2. $4\sec^2 2x\tan 2x$.
3. $4\sec^2 2x\tan 2x$.
4. $9\sec^2 3x\tan^2 3x$.

5. $\operatorname{cosec} x - x \operatorname{cosec} x \cot x$.

6. $2x \cot 2x - 2x^2 \operatorname{cosec}^2 2x$.

7. $\sec x \tan^2 x + \sec^3 x$.

8. $\frac{1}{2} \sin x (\cos x)^{-\frac{3}{2}}$.

9. $-6 \operatorname{cosec}^3 2x \cot 2x$.

10. $mx^{m-1} \tan^n x + nx^m \tan^{n-1} x \sec^2 x$.

11. $-mx^{-m-1} \sec^m x + mx^{-m} \sec^m x \tan x$.

12. $\frac{1}{2}x^{-\frac{1}{2}} \tan^2 x + 2x^{\frac{1}{2}} \tan x \sec^2 x$.

Examples V:

1. (i) $2, \dfrac{-2}{\sqrt{3}}, \frac{1}{4}$. (ii) $2, \dfrac{2}{\sqrt{3}}, \frac{1}{4}$. (iii) $2, \dfrac{-2}{\sqrt{3}}, \frac{1}{4}$.

(iv) $\dfrac{-2}{\sqrt{3}}, -2, \frac{3}{4}$. (v) $\dfrac{-2}{\sqrt{3}}, 2, \frac{3}{4}$. (vi) $\dfrac{2}{\sqrt{3}}, 2, \frac{3}{4}$.

2. $\tan^{-1} x + \dfrac{x}{1+x^2}$.

3. $\dfrac{2x}{\sqrt{(1-x^4)}}$.

4. $\cos^{-1} x - \dfrac{x}{\sqrt{(1-x^2)}}$.

5. $3x^2 \sin^{-1} 2x + \dfrac{2x^3}{\sqrt{(1-4x^2)}}$.

6. $2x \cos^{-1} x^2 - \dfrac{2x^3}{\sqrt{(1-x^4)}}$.

7. $\dfrac{2 \tan^{-1} x}{1+x^2}$.

8. $\dfrac{1}{x\sqrt{(x^2-1)}}$.

9. $\dfrac{-1}{x\sqrt{(x^2-1)}}$.

10. $\dfrac{-1}{1+x^2}$.

11. $\dfrac{2 \sin^{-1} x}{\sqrt{(1-x^2)}}$.

12. $(\cos^{-1} x)^3 - \dfrac{3x(\cos^{-1} x)^2}{\sqrt{(1-x^2)}}$.

13. $2x \sec^{-1} x + \dfrac{x}{\sqrt{(x^2-1)}}$.

14. $\operatorname{cosec}^{-1} x - \dfrac{1}{\sqrt{(x^2-1)}}$.

15. $2x(\sin^{-1} x)^2 + \dfrac{2x^2 \sin^{-1} x}{\sqrt{(1-x^2)}}$.

16. $\dfrac{-1}{(\sin^{-1} x)^2 \sqrt{(1-x^2)}}$.

Examples VI:

1. 6 per cent.

2. 2 per cent.

3. 0·2.

4. 0·00023.

CHAPTER III

Examples I:

1. $\dot{x} = -\pi \sin \pi t$, $\ddot{x} = -\pi^2 \cos \pi t$.
2. $\dot{x} = \frac{1}{2}\pi \cos \frac{1}{2}\pi t$, $\ddot{x} = -\frac{1}{4}\pi^2 \sin \frac{1}{2}\pi t$.
3. $\dot{x} = 5 - 64t$, $\ddot{x} = -64$.
4. $\dot{x} = 64t$, $\ddot{x} = 64$.
5. $\dot{x} = \sin \frac{1}{2}\pi t + \frac{1}{2}\pi t \cos \frac{1}{2}\pi t$, $\ddot{x} = \pi \cos \frac{1}{2}\pi t - \frac{1}{4}\pi^2 t \sin \frac{1}{2}\pi t$.
6. $\dot{x} = 2t \cos \frac{1}{2}\pi t - \frac{1}{2}\pi t^2 \sin \frac{1}{2}\pi t.$,
   $\ddot{x} = 2 \cos \frac{1}{2}\pi t - 2\pi t \sin \frac{1}{2}\pi t - \frac{1}{4}\pi^2 t^2 \cos \frac{1}{2}\pi t$.
7. $\dot{x} = 2t - 96$, $\ddot{x} = 2$.
8. $\dot{x} = \sin^2 \frac{1}{2}\pi t + \frac{1}{2}\pi t \sin \pi t$, $\ddot{x} = \pi \sin \pi t + \frac{1}{2}\pi^2 t \cos \pi t$.

Examples II:

2. 1.
3. -4.

Examples III:

1. 0.
2. 1.
3. -1 and 1.
4. -1, 0, and 1.
5. $n\pi + \frac{1}{2}\pi$, all integral n.
6. $\frac{1}{2}n\pi + \frac{1}{4}\pi$, all integral n.
7. $n\pi$, all integral n.
8. $2n\pi$, all integral n.

Examples IV:

4. (i) $x > \frac{1}{2}$.
   (ii) All x.
   (iii) $|x| > 1$.
   (iv) $x > 0$.
   (v) $-1 < x < 0$ and $x > 1$.
   (vi) $x < 1$ and $x > 4$.

Examples VI:

1. Minimum.
2. Minimum.
3. Minimum at $x = 1$, maximum at $x = -1$.
4. Minimum at $x = -1$ and 1, maximum at $x = 0$.
5. Minimum at $x = 2n\pi - \frac{1}{2}\pi$, maximum at $x = 2n\pi + \frac{1}{2}\pi$.
6. Minimum at $x = n\pi - \frac{1}{4}\pi$, maximum at $x = n\pi + \frac{1}{4}\pi$.

7. Minimum at $x = (2n+1)\pi$, maximum at $x = 2n\pi$.

8. Minimum at $x = 2(2n+1)\pi$, maximum at $x = 4n\pi$.

10. (i) Maximum at $x = -2$, inflexion at $x = 0$, minimum at $x = 2$.

(ii) Maximum at $x = -1$, inflexion at $x = -\frac{1}{2}$, minimum at $x = 0$.

11. Maximum at $x = 2n\pi + \frac{\pi}{4}, 2n\pi + \frac{3\pi}{4}, 2n\pi + \frac{6\pi}{4}$.

Minimum at $x = 2n\pi + \frac{2\pi}{4}, 2n\pi + \frac{5\pi}{4}, 2n\pi + \frac{7\pi}{4}$.

12. (i) Maximum at $x = 1$, minimum at $x = 2$.

(ii) (*a*) $x < 1$ and $x > 2$. (*b*) $1 < x < 2$.

Examples IX:

2. (i) $\xi = 1$. (ii) $\xi = 2$. (iii) $\xi = 1$.

Examples XI

1. $\frac{3}{2}$. 2. 0. 3. π. 4. 1. 5. $\frac{1}{8}$. 6. 1.

REVISION EXAMPLES I

1. (i) $x \cos x$. (ii) $6x - 10$. (iii) $\frac{-2x}{(1+x^2)^2}$.

2. (i) $\frac{4x - 5x^2}{2\sqrt{(1-x)}}$. (ii) $\sin 2x \cos 2x$. (iii) $\frac{1+2x-x^2}{(1+x^2)^2}$.

3. (i) $2x - \frac{4}{x^3}$. (ii) $2 \tan x \sec^2 x$.

4. (i) $2x - \frac{2}{x^3}$. (ii) $\frac{\cos x}{(1+\sin x)^2}$. (iii) $\frac{-x}{\sqrt{(a^2-x^2)}}$.

5. (i) $x^2(1+x)(3+5x)$. (ii) $2 \sin 4x$. (iii) $\frac{-2x}{\sqrt{(1-2x^2)}}$.

6. (i) $6x(1-x)$. (ii) $2\cos 2x \cos x - \sin 2x \sin x$. (iii) $\frac{x}{(1-x^2)^{\frac{3}{2}}}$.

7. $2x - y - 1 = 0$, $6x - 3y - 8 = 0$.

8. O, $2x - y = 0$; A, $x + y - 1 = 0$; B, $2x - y - 4 = 0$. Intersection $\left(\frac{5}{3}, -\frac{2}{3}\right)$.

9. Maximum at $x = 0$, minimum at $x = 2$. Parallel tangent at $x = -1$.

10. $4x - 2y + 5 = 0$. Point $\left(\frac{2}{3}, \frac{41}{27}\right)$.

11. $\left(-\sqrt{\frac{2}{3}}, -\frac{1}{3}\sqrt{\frac{2}{3}}\right)$, $(0, 0)$, $\left(\sqrt{\frac{2}{3}}, \frac{1}{3}\sqrt{\frac{2}{3}}\right)$.

12. $5x - y - 1 = 0$, $(2, 4)$, $(4, -8)$.

13. Gradient $4h$, tangent $y - k = 4h(x - h)$. Tangents through origin $y = \pm 12x$.

14. $x = \frac{5}{6}$ and $-\frac{3}{2}$.

15. $a = 2$, $b = -9$, $c = 12$, $d = 0$.

16. Maximum value 9 at $x = -1$; minimum value -7 at $x = 1$ and at $x = -3$.

17. Maximum value 4 at $x = -2$; minimum value 0 at $x = 2$. Tangent $3x + 2y - 4 = 0$.

18. $(-1, 27)$ and $(2, 0)$.

19. $(1, 6)$ and $(3, 2)$.

20. Gradient zero, minimum.

21. $(1, 2)$ and $(3, -2)$. Three positive roots, one between 0 and 1, one between 1 and 3, and one greater than 3.

22. Velocity $= 2pt + 3qt^2$, acceleration $= 2p + 6qt$, $p = 12$, $q = -1$.

24. Velocity $= 0$, acceleration $= -4$. $1\frac{1}{2}$ seconds, 18 feet.

25. Velocity 16 ft./sec. Acceleration 14 ft./sec.^{2}

26. Velocity $= t \cos t$. Acceleration $= \cos t - t \sin t$.

27. $\frac{9}{2}$ sq. in., 4 sq. in.

28. 32 ft.

29. $x = \frac{4a}{3}$. Time $\frac{3a}{4}$ hours.

30. $x = 13{\cdot}7$.

31. 4500 cu. ft.

32. Volume $7\frac{7}{8}$ cu. ft. Length $4{\cdot}09$ ft.

33. 48 sq. ft.

34. Area $= \frac{1}{18}(l^2 - 2lx + 2x^2)$. $x = \frac{l}{2}$.

37. Area $= \frac{1}{16}x^2 + \frac{1}{18}(l-x)^2$. Minimum.

39. $\frac{2}{3}h$ per cent.

40. $\dfrac{3}{2\pi}$ cu. in. per sec.

41. Shortened, $\frac{5}{72}$ per cent.

42. (i) $0{\cdot}15\,\pi$ cu. in. per sec. (ii) $0{\cdot}16\,\pi$ sq. in. per sec.

43. (i) $0{\cdot}00008$ in./sec. (ii) $0{\cdot}025$ cu. in./sec.

44. $\dfrac{dr}{dt} = -3$ in. per hour. 24π sq. in. per hour.

45. $\frac{1}{3}h$ per cent.

CHAPTER IV

Examples II:

2. $A + \frac{1}{4}x^4,\ A - \dfrac{1}{2x^2},\ A + \frac{1}{6}x^6$.

3. $A + \frac{1}{2}\sin 2x,\ A - 2\cos\frac{1}{2}x,\ A + x + \sin x$.

5. $8\frac{2}{3}$, 0, 1, 66.

6. 1, $\frac{1}{2}$, π, $\frac{1}{4}\pi$.

7. $\frac{1}{3}, \frac{1}{6}, \dfrac{21352}{405}$.

8. $\frac{1}{2}\pi$.

CHAPTER V

Examples I:

1. $C - \frac{1}{2}\cos 2x$.
2. $C + \frac{1}{3}\sin 3x$.
3. $C - 2\cos\frac{1}{2}x$.
4. $C + \frac{1}{4}\tan 4x$.
5. $C + 2\sec\frac{1}{2}x$.
6. $C + \frac{1}{3}(x+1)^3$.
7. $C + \frac{1}{4}(x+3)^4$.
8. $C + \frac{1}{5}(x+5)^5$.
9. $C - \frac{1}{3}(x+1)^{-3}$.
10. $C - \frac{1}{4}(x-2)^{-4}$.
11. $C - \frac{1}{2}(x-1)^{-2}$.
12. $C - (x+5)^{-1}$.
13. $C + \frac{1}{6}(2x+1)^3$.
14. $C + \frac{1}{20}(5x-3)^4$.

15. $C+\frac{2}{5}(\frac{1}{2}x+7)^5$.

16. $C-\frac{1}{2}(2x+1)^{-1}$.

17. $C-\frac{1}{8}(4x-3)^{-2}$.

18. $C-(\frac{1}{3}x-1)^{-3}$.

19. $C+\frac{2}{3}(x+1)^{\frac{3}{2}}$.

20. $C+\frac{2}{5}(x-1)^{\frac{5}{2}}$.

21. $C+2(x+1)^{\frac{1}{2}}$.

22. $C+(2x+1)^{\frac{1}{2}}$.

23. $C-(2x-3)^{-\frac{1}{2}}$.

24. $C+\frac{1}{2}(3x+1)^{\frac{2}{3}}$.

25. $C-\frac{1}{2}(x^2+1)^{-1}$.

26. $C-\frac{1}{9}(x^3+1)^{-3}$.

27. $C-\frac{1}{15}(3x^5+1)^{-1}$.

28. $C+\frac{1}{20}(1-5x^2)^{-2}$.

29. $C-\frac{1}{6}\cos^6 x$.

30. $C+\frac{1}{3}\cos^3 x-\cos x$.

31. $C+\frac{1}{3}\sin^3 x-\frac{1}{5}\sin^5 x$.

32. $C+\frac{1}{3}\tan 3x-x$.

33. $C-\frac{1}{2}(1+\tan x)^{-2}$.

34. $C+\sec x+\frac{1}{2}\tan^2 x$.

35. $C+\tan x+\frac{1}{3}\tan^3 x$.

36. $C+\frac{1}{3}\tan^3 x$.

Examples II:

1. $C+\dfrac{1}{16}\tan^{-1}\dfrac{x}{2}+\dfrac{1}{8}x(x^2+4)^{-1}$.
2. $C+\frac{1}{2}\sin^{-1}x-\frac{1}{2}x(1-x^2)^{\frac{1}{2}}$.
3. $C+x-\tan^{-1}x$.
4. $C+\frac{1}{2}\sin^{-1}x+\frac{1}{2}x(1-x^2)^{\frac{1}{2}}$.
5. $C+\sin^{-1}x-(1-x^2)^{\frac{1}{2}}$.
6. $C+\frac{1}{8}\sin^{-1}x+\frac{1}{8}x(2x^2-1)(1-x^2)^{\frac{1}{2}}$.
7. $C-\frac{2}{3}(2+x)(1-x)^{\frac{1}{2}}$.
8. $C-\frac{2}{3}(18+x)(9-x)^{\frac{1}{2}}$.
9. $C-\frac{2}{15}(2+3x)(1-x)^{\frac{3}{2}}$.
10. $C+\frac{2}{105}(15x^2+48x+128)(x-4)^{\frac{3}{2}}$.

Examples III:

1. $\frac{1}{3}$. 2. $\frac{2}{3}$. 3. $\dfrac{\pi}{8}+\dfrac{1}{4}$. 4. $\frac{1}{4}\pi$. 5. $\dfrac{\pi}{12}+\dfrac{\sqrt{3}}{8}$.

6. $\frac{1}{3}$. 7. $\frac{4}{3}$. 8. $\dfrac{8}{15}-\dfrac{49}{160}\sqrt{3}$. 9. $\frac{2}{3}$. 10. $\frac{14}{3}$.

Examples IV:

1. 1. 2. $\frac{2}{3}$. 3. $\frac{2}{7}$.

Examples V:

1. $C+x\sin x+\cos x$.
2. $C+x^2\sin x+2x\cos x-2\sin x$.
3. $C+x\sec^2 x-\tan x$.
4. $C+\frac{1}{2}x^2\tan^{-1}x-\frac{1}{2}x+\frac{1}{2}\tan^{-1}x$.
5. $\frac{1}{2}\pi-1$.
6. $\dfrac{\pi}{\sqrt{3}}-1-\dfrac{\pi^2}{18}$.
7. $\pi^2+2\pi-2$.
8. $6-2\pi+\frac{1}{4}\pi^2$.

Examples VI:

1. $C-\frac{1}{12}\cos 6x-\frac{1}{8}\cos 4x$.
2. $C+\frac{1}{2}x+\frac{1}{4}\sin 2x$.
3. $C+\frac{1}{2}x-\frac{1}{16}\sin 8x$.
4. $C+\frac{1}{4}\sin 2x-\frac{1}{12}\sin 6x$.
5. $-\frac{1}{15}$.
6. $\frac{4}{3}$.
7. $\frac{2}{3}$.
8. $\dfrac{3\pi}{16}$.

Examples VII:

1. $C-\dfrac{1}{6}\sin^5 x\cos x-\dfrac{5}{24}\sin^3 x\cos x-\dfrac{5}{16}\sin x\cos x+\dfrac{5}{16}x$.
2. $C+\dfrac{1}{7}\sin x\cos^6 x+\dfrac{6}{35}\sin x\cos^4 x+\dfrac{8}{35}\sin x\cos^2 x+\dfrac{16}{35}\sin x$.
3. $$C-\frac{1}{9}\sin^8 x\cos x-\frac{8}{63}\sin^6 x\cos x-\frac{16}{105}\sin^4 x\cos x$$
$$-\frac{64}{315}\sin^2 x\cos x-\frac{128}{315}\cos x.$$
4. $\dfrac{35\pi}{256}$.
5. $\dfrac{16}{35}$.
6. $\dfrac{63\pi}{512}$.
7. $\dfrac{(2n-1)(2n-3)\ldots 3.1}{2n(2n-2)\ldots 4.2}\cdot\dfrac{\pi}{2}$.
8. $\dfrac{2n(2n-2)\ldots 4.2}{(2n+1)(2n-1)\ldots 5.3}$.
9. $\dfrac{2n(2n-2)\ldots 4.2}{(2n+1)(2n-1)\ldots 5.3}$.

Examples VIII:

1. $\dfrac{16}{1155}$.
2. $\dfrac{2}{63}$.
3. $\dfrac{m!\,n!}{(m+n+1)!}$.
4. $\dfrac{(2m-1)^2(2m-3)^2\ldots 3^2}{2^{2m}(2m)!}\cdot\dfrac{\pi}{2}$.
5. $\dfrac{(m!)^2}{2(2m+1)!}$.
6. $2.\dfrac{12.10.\ldots.2}{19.17.\ldots.7}$.
7. $\dfrac{7\pi}{2^{10}}$.

CHAPTER VI

Examples I:

1. π. 2. $\frac{3}{2}\pi$. 3. 11π.

Examples II:

1. (*a*) 0·335, (*b*) 0·333. 2. (*a*) 60·1, (*b*) 60·0.
3. (*a*) 8·68, (*b*) 8·67. 4. (*a*) 0, (*b*) 0.
5. (*a*) 0·680, (*b*) 0·667. 6. (*a*) −0·248, (*b*) −0·250.
7. (*a*) 648, (*b*) 638. 8. (*a*) 0·694, (*b*) 0·693.
9. (*a*) 0·322, (*b*) 0·322.

Examples III:

1. $\frac{10}{3}a^2$. 2. $\frac{8}{3}l^2$. 3. $\frac{3}{2}a$.
4. (i) $\dfrac{3a^2}{49}(31+6\sqrt{2})$. (ii) $\dfrac{a^2}{245}(325+6\sqrt{2})$.

REVISION EXAMPLES II

1. $\frac{1}{6}(2x-3)^3$, $\frac{6}{5}x^{\frac{5}{3}}$.
2. $\frac{2}{3}$, $\frac{2}{5}$, $\frac{1}{2}$.
3. $\frac{1}{5}x^5+2x-\dfrac{1}{3x^3}$, $\dfrac{8}{3}$.
4. (i) $\frac{1}{3}x^3-\frac{1}{2}x^4+\frac{1}{5}x^5$. (ii) $\frac{1}{2}x-\frac{1}{4}\sin 2x$. (iii) $\frac{2}{3}$.
5. (i) $\frac{1}{4}(x+1)^4$. (ii) $\frac{1}{2}x+\frac{1}{12}\sin 6x$. (iii) $\frac{1}{8}\sqrt{3}$.
6. (i) $\frac{1}{2}x-\frac{1}{4}\sin 2x$. (ii) $-\dfrac{1}{x}-\dfrac{1}{3}x^3$. (iii) $-\frac{2}{3}\cos^3 x$.
7. $10a$.
8. $y=\frac{3}{2}x^2-2x+4$, $2y-x-8=0$.
9. $y=x^4-\frac{1}{2}x^2+1$, minima at $x=\pm\frac{1}{2}$.
10. $y=x-\dfrac{x^3}{6}+\dfrac{10}{3}$, $(-2,\frac{8}{3})$, $x+y-6=0$ at $(2,4)$, $x+y-\frac{2}{3}=0$ at $(-2,\frac{8}{3})$.
11. $\frac{1}{6}$, $\frac{1}{24}$. 12. $y=x+\dfrac{1}{x}+1$, $(1,3)$.

13. $8\frac{2}{3}$, $4\frac{2}{3}$.

14. 10 ft. per sec., $31\frac{1}{3}$ ft.

15. 27 ft., $85\frac{1}{3}$ ft. after 8 sec.

16. 9 ft. per sec., 3 ft. per sec. per sec.

17. 3 ft. per sec. per sec. when $t = 2$, -1 ft. per sec. per sec. when $t = 4$. $AB = 11\frac{1}{3}$ ft. At rest when $t = 1, 6$.

18. 50 ft. per sec., $1166\frac{2}{3}$ ft.

19. 4 ft. per sec., $4\frac{1}{3}$ ft.

20. $v = \pm 5$, max. $x = \frac{5}{2}$.

21. -12 cm. per sec. per sec., $60\frac{5}{12}$ cm., $19\frac{7}{12}$ cm.

22. 31·25 ft. per sec., $520\frac{5}{6}$ ft.

23. $t = 5$; $5\frac{1}{3}$ ft.

24. $t = 2, 5$; $4\frac{1}{2}$ ft.

25. 14 ft. per sec. per sec., 25 ft.

26. 15.

27. $\frac{8}{3}$, $(\frac{3}{2}, \frac{6}{5})$.

28. $6\frac{3}{4}$, $\frac{1}{5}$.

29. (i) $\frac{1}{4}$, (ii) $\frac{1}{4}$.

30. $(\frac{9}{5}, \frac{54}{35})$.

31. $\frac{7}{3}$, $(\frac{45}{28}, \frac{93}{70})$.

32. $(\frac{9}{8}, \frac{18}{5})$.

33. $\frac{12}{7}$.

34. 36, $(0, -\frac{18}{5})$.

35. $\frac{16}{3}$, $(\frac{3}{2}, 0)$.

36. 9, $(\frac{3}{2}, \frac{9}{5})$.

37. $\dfrac{4a}{3\pi}$,

38. 32π, $\frac{8}{3}$.

39. (i) 8π, (ii) $\frac{9}{8}$.

40. $\frac{32}{5}\pi$, $\frac{31}{24}$.

41. $\frac{4}{3}\pi$, $\frac{6}{5}$.

42. (i) $\frac{128}{5}\pi$, (ii) 16π.

44. 4, $\frac{128}{7}\pi$.

45. $\frac{1}{3}\pi r^2 h$, $\frac{3}{4}h$.

46. $\frac{56}{3}\pi$.

47. 125π.

48. $\frac{64}{3}\pi$.

49. $\frac{4}{3}\pi ab^2$, $\frac{3}{8}a$.

50. $\frac{112}{3}\pi$.

51. $\dfrac{\pi^2 a^2}{4}$, $\dfrac{\pi}{4} + \dfrac{1}{\pi}$.

52. $\frac{218}{3}$.

53. 2·55.

54. 0·292.

55. 0·84.

56. 710 sq. ft.

58. 2·7 miles.

59. $a = \frac{24}{45}$, $b = \frac{64}{45}$, $c = \frac{14}{45}$.

60. 0·385.

61. Exact method 0·867. Simpson's rule 0·865.

62. 50·3.

63. 342.

INDEX